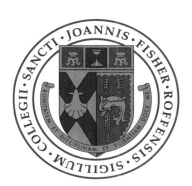

Contributors

Jack Bookman   Department of Mathematics, Duke University, Durham, NC 27708

Alphonse Buccino   Department of Mathematics (Emeritus), University of Georgia, Athens, GA 30602

Paul Davis   Department of Mathematical Sciences, Worcester Polytechnic Institute, Worcester, MA 01609

Wade Ellis   Department of Mathematics, West Valley College, Saratoga, CA 95070

Susan Ganter   Department of Mathematical Sciences, Clemson University, Clemson, SC 29634

Melvin George   Department of Mathematics (Emeritus), University of Missouri, Columbia, MO 65211

Sheldon Gordon   Department of Mathematics, State University of New York at Farmingdale, Farmingdale, NY 11735

Harvey Keynes   IT Center for Educational Programs, University of Minnesota, Minneapolis, MN 55455

James Lightbourne   Division of Undergraduate Education, National Science Foundation, Arlington, VA 22230

William McCallum   Department of Mathematics, University of Arizona, Tucson, AZ 85721

Dan O'Loughlin   IT Center for Educational Programs, University of Minnesota, Minneapolis, MN 55455

Andrea Olson   IT Center for Educational Programs, University of Minnesota, Minneapolis, MN 55455

Douglas Shaw   IT Center for Educational Programs, University of Minnesota, Minneapolis, MN 55455

David Smith   Department of Mathematics, Duke University, Durham, NC 27708

Preface

Calculus Reform. Or, as many would prefer, calculus renewal. These are terms that, for better or worse, have become a part of the vocabulary in mathematics departments across the country. The movement to change the nature of the calculus course at the undergraduate and secondary levels has sparked discussion and controversy in ways as diverse as the actual changes. Such interactions range from "coffee pot conversations" to university curriculum committee agendas to special sessions on calculus renewal at regional and national conferences. But what is the significance of these activities? Where have we been and where are we going with calculus and, more importantly, the entire scope of undergraduate mathematics education?

In April 1996, I received a fellowship from the American Educational Research Association (AERA) and the National Science Foundation (NSF). This fellowship afforded me the opportunity to work in residence at NSF on a number of evaluation projects, including the national impact of the calculus reform movement since 1988. That project resulted in countless communications with the mathematics community and others about the status of calculus as a course in isolation and as a significant player in the overall undergraduate mathematics and science experience for students (and faculty).

While at NSF (and through a second NSF grant received while at the American Association for Higher Education), I also was part of an evaluation project for the Institution-wide Reform (IR) program. The first grants in the IR program were awarded by the NSF Division of Undergraduate Education in 1996 to colleges and universities working to integrate science and mathematics across disciplinary boundaries, including projects that incorporated the arts, social sciences, and teacher preparation. The primary goal for such projects was to create an environment for learning science that encourages science literacy for all students at the undergraduate level.

My NSF-sponsored activities also included the charge to use knowledge gained from the calculus and IR programs in the development of questions that need further investigation. In response to this charge, as well as the outcomes of more than 10 years of changing and re-changing the calculus course, I propose that it is very important to consider the following:

1. What exactly have we learned—about calculus, about undergraduate mathematics education, and about the interaction of the discipline of mathematics with other courses in the sciences and beyond?
2. How can this knowledge be applied in positive ways to the organization and delivery of undergraduate mathematics courses while maintaining the mathematical integrity of these courses?
3. What questions still need to be asked and answered, through carefully conducted research, to guide future changes?

It is the third of these questions that motivates this publication. Specifically, the first several years of the calculus reform movement were characterized by a whirlwind of ideas about the organization of the course and the associated curriculum. Soon after came a dissemination phase, where other institutions eager to participate and improve their own courses "jumped on the bandwagon." Throughout, skeptics of the changes—and even many of the "reformers" themselves—began more and more to question the justification for these changes and the implications for faculty development, recruitment of mathematics majors, and the place for mathematics within the undergraduate curriculum. Even though many mathematicians conceded that change was necessary, they questioned the fact that the current changes were being made without first carefully examining—as a community, rather than individually by project—the overriding objectives and goals for students enrolled in first-year collegiate mathematics courses. Many claimed that the development of such goals is critical to the discipline of mathematics because this course will in fact be the last formal mathematical training for a large majority of our students.

Given these valid concerns, it is important that the mathematics community develop, research, and form workable solutions for many difficult questions. The purpose of this publication is to pose and discuss a small subset of these questions, including the following:

• What are the contributions that mathematics can—and should—make in the world of science, engineering, and technology? How can we better nurture our relationships with colleagues in these other disciplines? (George, Lightbourne)
• What mathematical skills and knowledge should an undergraduate student have after completing a first-year collegiate mathematics course? (McCallum)

- What has been learned about the cognitive processes involved in learning mathematics through the extensive research conducted in this area during the past decade? How can we use this research to better inform the development of undergraduate mathematics courses? (Smith)
- How does mathematics fit into the broader context of student learning; e.g., what are the life skills that our students will need to succeed in the workplace? (Davis)
- What is the appropriate role of technology in the teaching and learning of mathematics? (Ellis)
- How does a first-year collegiate mathematics course contribute to the overall mathematics education of our students? How can the changes made in this course be used to improve other mathematics courses? (Gordon)
- What is the appropriate role of colleges and universities in supporting curricular change? How can administrators provide an environment that is conducive to change and enables faculty to develop the necessary skills for supporting such change? (Keynes *et al.*)
- What are the appropriate mechanisms by which the mathematics community can evaluate progress and thereby better inform continuing change? What changes ultimately help our students to better understand mathematics and to have an appreciation for the importance of mathematics in our society? (Bookman, Buccino)

It is my hope that the papers contained herein will spark a renewed interest in the endeavor embarked upon over 10 years ago when the first calculus grants were awarded by NSF. However, it is certain that the next decade of calculus renewal will be tackled by a mathematics community that has learned from experience and has considered questions such as those presented here. We have come a long way in the development of an environment in undergraduate mathematics education that is conducive to positive change. The true success of these efforts will be measured by the fervor with which the mathematics community builds on these efforts and defines the appropriate place for mathematics in the undergraduate curriculum of the 21st century.

Susan L. Ganter
Department of Mathematical Sciences
Clemson University

Contents

CHAPTER 1

Calculus Renewal in the Context of Undergraduate SMET Education

Melvin D. George

The calculus reform movement occurred in the context of other concerns and developments related generally to science, mathematics, engineering, and technology (SMET) education. It is important to understand that general context, as well as to consider specific issues that have recently become clear in that context, which suggest agendas for the next developments in undergraduate mathematics education.

The National Science Foundation (NSF) has been the largest federal source of funding for projects related to improving undergraduate SMET education generally; and, indeed, many calculus reform projects were supported by the NSF's undergraduate education division in cooperation with several other NSF divisions, including the mathematical sciences. To understand the national background for the calculus reform movement, therefore, it is helpful to use the NSF as a frame of reference.

1. SOME HISTORY

The NSF has, over a long period of time, supported graduate education; indeed, NSFs Graduate Fellowship Program was one of the components of the original NSF budget in FY1952. More recently, and reinforced in 1983s *A Nation at Risk*,[1] there has been heavy emphasis within NSF on K–12 education, on the development of educational materials and curricula, on professional development for teachers, and on other programs designed to improve elementary and secondary education in science and mathematics.

Prior to the 1980s, NSF had a relatively small program of support for undergraduate SMET education, largely focused on individual grants for equip-ment or other projects designed to improve curricula. But even this relatively modest effort was put in jeopardy when the Reagan administration moved, as it came into office, to eliminate all funding for undergraduate education from the portfolio of the NSF. Consequently, funding support for undergraduate education improvements vanished. That circumstance became the motivation for the seminal NSF review of undergraduate science education, *Undergraduate Science, Mathematics and Engineering Education*, commonly known as the Neal Report after the chairman of the NSB committee that wrote the document, Homer Neal of the University of Michigan.[2]

The major emphasis of the report, issued in 1986, was on the preparation of majors in SMET fields (though technology played virtually no role in the Neal Report), and it specifically urged the restoration and expansion of the program of individual grants for instrumentation and laboratory improvement that had been a hallmark of NSF efforts. The report was quite successful; the instrumentation program was maintained and others at the undergraduate level added. Indeed, the instrumentation and laboratory improvement program has for some time been the largest program within the entire NSF as measured by the number of proposals. The size of this program reflects the historic attention paid by the NSF, even in its funding of *educational* improvements, to individual faculty members and individual projects. Given the research model on which NSF is based, this is certainly not surprising.

The Neal Report itself, however, included a strong recommendation that a separate division of NSF continue to be responsible for undergraduate education (rather than assigning all educational functions to the research directorates) and that a new program of "comprehensive projects" be established within that division (now the Division of Undergraduate Education), moving away from the model of exclusive reliance on the individual investigators model. The proposal to undertake support of something akin to *institutional* reform is the only recommendation in the Neal Report that was never implemented in the late 1980s, but it reflects the belief already then developing within NSF that real and lasting change would not come about solely through support of individual faculty members to do individual improvement projects, no matter how valuable and worthwhile those projects might be.

2. DEVELOPMENTS SINCE THE NEAL REPORT

A lot happened in the decade between 1986 and 1996, when the Education and Human Resources Directorate of NSF issued another report reviewing the status of undergraduate SMET education, *Shaping the Future*.[3] A growing understanding of the importance of the linkages between K–12 education (which continues to have very high priority nationally) and the undergraduate level developed. A major and obvious example is that K–12 teacher preparation programs conducted at the undergraduate level must be included in any significant and long-lasting program of elementary and secondary improvement.

There was also a growing appreciation that the U.S. work force is changing. We have seen a shrinking number of available high-paying jobs requiring low skills; consequently, a much larger fraction of the nation's population will need some understanding of science, mathematics, engineering, or technology in order to succeed in the workplace (for further discussion, see Davis, Chapter 4). Related to this is a growing concern about the problem of what some have called "scientific literacy," the level of general understanding of what science is and is not, of what science can do and cannot do, of the nature of scientific truth, and of how decisions should take science into account. Some of the factors leading to this concern certainly include increasingly sophisticated medical questions, issues related to genetic engineering, and major investment policy decisions at the federal level such as the superconducting supercollider and star wars technology.

It became increasingly clear that undergraduate SMET education had to be viewed more broadly than the preparation of science majors and future professional scientists and mathematicians. Those responsible for undergraduate SMET education had to be increasingly knowledgeable about the workplace, and increasingly concerned about higher education's role in the preparation of K–12 teachers and in developing a populace with a better broad understanding of the basic principles of science. There was a growing feeling that the problems of SMET undergraduate education were systemic and that solutions must include efforts that are broader than improving individual courses and individual laboratories in individual institutions.

The calculus reform movement was an early and very significant move by NSF along the continuum from individual projects to systemic programs, moving to reshape a basic curricular element across the country. The effort surely reflected a concern that a broader sense of responsibility—moving beyond individual faculty members and toward more departmental and institutional commitments—was necessary. There was a parallel growth within the Education and Human Resources directorate of the state systemic initiatives, and private foundations such as the Howard Hughes Medical Institute developed institutionally based programs to try to encourage lasting and sustained improvement in undergraduate SMET education.

3. SHAPING THE FUTURE

Shaping the Future and its counterpart *From Analysis to Action,*[4] sponsored by the National Research Council in association with NSF, together with similar reports and efforts under way across the country during the decade, came to similar conclusions. Undergraduate SMET education must be concerned about *all* students, not just the few who have chosen to major in these fields, and it must stress learning the methods and processes of science and mathematics, not just the memorization of facts and the practice of skills. This vision of undergraduate SMET education has seemed to necessitate a parallel broadening of the methods used to bring about such change. When you give attention to all students, for example, the curricular, staffing, and space problems that result are much broader and much more institutional than are decisions about what and how to teach the few select who intend ultimately to get Ph.D. degrees in these fields. To try to change the emphasis and pedagogical style of calculus across the country, for example, presents quite a different kind of structural challenge from improving a junior organic chemistry course for majors.

As these ideas have developed on the national stage since 1986, the Division of Undergraduate Education at NSF has implemented an institutional reform program as suggested by the Neal Report, reflecting the conclusion that if real, sustained improvement is to occur in undergraduate SMET education—the kind of improvement that will lead to all students in all institutions having access to SMET courses and curricula that are engaging and connected, and that acquaint the students not only with the facts of science but also with its methods and processes—then a lot has to happen in institutions. There must be changes in (to name a few):

- Curricula
- Reward systems for faculty
- The kind of space used for instruction in science and mathematics
- Professional development of faculty
- Meaningful connections with elementary and secondary education
- Technology and materials
- How future faculty are trained during Ph.D. programs about teaching and learning in SMET

4. NEW AGENDA ITEMS

Out of all of these developments, discussions, and programmatic changes on the national stage have come some new questions and new agenda items that apply to mathematics as to the other SMET fields. As mathematicians build on the calculus reform movement (a title soon to be replaced, one hopes, by calculus *renewal*, as the word "reform" has pejorative implications that cause unnecessary problems), what are some of these general issues that developments in the field ought to address?

4.1. Research in Learning

One of the emphases of *Shaping the Future*[3] is the need for more research focusing on how young people learn science and mathematics and for more use of the considerable research that already exists in decisions about changing courses, curricula, and classroom styles. In a subsequent chapter, Smith points out that the leaders of the calculus reform movement had certain commitments to ways of doing calculus that seemed intuitively superior, but the research base required to justify these decisions was not yet in place (see Chapter 3). Thanks to advances in cognitive sciences and other fields, much of that research has now been done and the foundations of our understanding of learning in science and mathematics are much stronger than even 10 years ago.

Best Uses of Technology. But there are still many difficult questions. A large number of those questions revolve around the best uses of technology. Some institutions have carefully thought about intended results of technological improvement in various course and curricular structures and have tried to assess the achievement of those objectives. But, although there are some who would disagree, it would appear that there is not yet a clear understanding of how best to use technology to achieve certain learning objectives. And there are certainly too many instances in which large amounts of money have been spent on technology that has produced little measurable (or at least, little *measured*) improvement in learning (for further discussion, see Ellis, Chapter 5).

Expanded General and Field-Specific Learning Research. There have been advances in general learning research as well as in field-specific under-standings of how people learn, but both must be greatly expanded. Both theoretical and field-based longitudinal studies should be supported, both in general learning and in aspects of learning specific to individual SMET fields. Studies such as those reported by Bookman (Chapter 7) are certainly an important beginning, and the recent budget request of the Clinton administration for substantial funding for long-term research in education is another step in the right direction, provided a suitable research agenda is developed for the use of these funds. However, even with the substantial amount of research that has already been done, few SMET faculty members and few SMET departments actually know about and *use* that research in designing courses and curricular experiences.

Engaging Faculty in Learning Research Results. There is a related issue that is critical to future efforts to build sustained improvement in undergraduate SMET education: SMET faculty must be engaged in the intellectually challenging question of how learning occurs best in their fields. It seems strange that people who are inherently intrigued by problems in their fields do not seem intellectually engaged in the question of how people learn those fields. Institutions could do a great deal to inculcate a sense of intellectual excitement and intellectual challenge associated with questions of mathematics learning, for example. The so-called "math wars" in California and elsewhere have produced far more heat than light about the difficult and important questions associated with mathematics education. College campuses should be places in which these questions are discussed, careful research is done, and results are used in intellectually viable ways to advance student learning.

A major agenda item for the next stage in the development of mathematics improvement at the undergraduate level is this question; "What is known and what additional knowledge is needed about how students actually acquire understanding and skill in mathematics?" There is also a subsidiary question, namely, "Once the first question is answered, how can faculty members be acquainted with the answer and persuaded of its value?" These are certainly very complicated questions, and the way in which they are posed may sound overly simplistic to many readers. But mathematicians as a whole need to engage in more thoughtful discussion and more directed action in these areas, much as we would attack a research problem in mathematics itself.

4.2. Assessment

Certainly related to the question of research about learning is the issue of assessment. The development of statewide assessment tests in mathematics, for example, has sharpened the question of whether or not we really know what we want to assess and how to assess it. More attention needs to be paid to the relationship between assessment and educational goals. In addition, it is critical to understand the profound influence of national tests such as SAT, ACT, and the Graduate Record Exam on what is taught, what is studied, and the expectations of students, parents, and faculty about what is important and what is not important.

Not enough attention has been paid to the issues of assessment in reports such as *Shaping the Future*.[3] Faculty members receive little help with assessment methodologies, for example. One particularly acute problem is that assessment is often defined as questions to which students should supply answers, the relative worth of which are then judged and graded by the professor. Perhaps a more significant indicator of the depth of understanding of a field is the extent to which a student can *generate* questions that are meaningful, interesting, and potentially important. However, a student's ability to create good questions is rarely assessed.

There is also a good bit of evidence that asking students to carry out open-ended problem-solving exercises is an important way of helping students develop higher-order thinking skills. But making such assignments poses particular assessment problems for faculty, and so, even if they are thought of by an individual professor, such assignments may be rejected because of the professor's unwillingness to deal with the resulting assessment challenges. Therefore, in the next decade of mathematics improvement—building on the initial efforts in calculus—it is important that mathematicians as a whole consider what is important to assess and how it might be assessed, including the question of how faculty can learn about various assessment methodologies and techniques and use them effectively to achieve educational goals (for a related discussion, see Buccino, Chapter 9).

4.3. Connections

There are a number of significant issues relating to connections—connections between what faculty are teaching and the "real world" that students are concerned about, connections among fields and subfields. Do we—should we—help students understand relationships between calculus and abstract algebra, between calculus and economics? Is it our function to teach "only" mathematics without any sense of where the mathematics came from, what it is "useful" for, and how it is related to all the other things the student is learning? In general, how do we help students integrate not only the individual parts of a rapidly increasing body of SMET material but also connect that body with learning in other fields? (for discussion, see McCallum, Chapter 2). One of the things that is known, as a result of increased understanding, is that the brain is not a tabula rasa, to be written on indelibly by whatever the professor says; it is an active organism that must establish connecting pathways by which new information becomes associated with previous knowledge if that information is to be available and usable for a long period of time. Most faculty don't understand well enough how to help the brain do that and do not always accept the benefits of spending time at making those connections, when doing so seems to be at the expense of proving just one more theorem.

4.4. Curricular Issues

Certainly related to the general issue of connections is a whole array of curricular questions, and there are particular questions about "connections" for mathematics. Now that calculus renewal is so far along, how do the new calculus courses relate to the rest of the mathematics curriculum, both for nonmajors and for majors? (The experience with the Ph.D. program at the University of Rochester certainly alerted mathematicians throughout the country to the dangers of treating their curriculum as virtually disjoint from the curricular needs and interests of other SMET departments.) How can mathematicians design a workable curriculum in mathematics that will make possible the superb training of majors while still dealing effectively with nonmajors? What about prerequisites to calculus (see Gordon, Chapter 6)? Would it be better to do general mathematical education in courses at the junior level, for example, related to the various majors of the students, than to try to do them as "introductory" (really "terminal") course experiences?

A curricular issue that should have higher priority is how to deal with students who switch majors. The undergraduate curricular structure seems to assume that a student, at the moment he or she arrives on campus, is immutably fixed on either being a math major or not being a math major—or at least on being a SMET major or not being a SMET major. There may be a curriculum structure for potential majors and another one for nonmajors, but isn't it strange that these don't easily allow for the possibility that a student who does not intend originally to major in mathematics (or in SMET generally) might become so turned on by a course that he or she changes direction? Or is that, sadly, such a rare occurrence as to have probability close to zero? And what about a mathematics curriculum for prospective elementary and middle school teachers? This topic will be considered in the next section of this chapter, but it is an important issue to think about when building a curriculum in mathematics.

And where in the curriculum do we deal with students' beliefs about mathematics? With the development of nonmajors courses that stress the structure of a particular SMET field and its ethical implications comes the risk that only nonmajors will acquire an understanding of what the field really is, what is the nature of "truth" in the field, and what ethical issues arise within the field. It seems strange and dangerous not to acquaint our majors with these same concerns. But how do we deal with this as a curricular issue?

4.5. Preparation of K–12 Teachers of Mathematics

Finally, but certainly not of least importance, is the issue of the preparation of elementary and secondary teachers. Despite the increasing amount of rhetoric that is given to teacher preparation programs in colleges and universities, the overall problems in education will not be solved until we figure out a way to produce stronger teachers of mathematics who care about the mathematical learning of all students and know how to achieve that.

The answer is not simply to require all prospective mathematics teachers to have a major in mathematics, as it is not certain that a standard mathematics major provides the right mathematical understanding for one preparing to teach children. Mathematics baccalaureate majors do not necessarily have the ability to engage children in mathematics learning that K–12 teachers should have. And, sadly to say, it may be that they will not learn much about this by observing the teaching that they receive in their undergraduate mathematics courses (for further discussion, see Keynes *et al.*, Chapter 8).

Institutions simply have to find a way to build meaningful programs that are strong mathematically but include real attention to how to engage students in learning. The Mathematical Sciences Education Board prepared a report[5] about teacher preparation in mathematics, which deserves wider attention and discussion in the field. Basing efforts to build professional development programs in *mathematics* for current teachers of mathematics on engaging the teachers in important questions that arise in the classroom seems very promising. Much deep mathematics, for example, can probably be taught to current teachers by examining incorrect answers given by students on homework or tests and helping the teachers deal with why and how those students got those answers and what they reveal mathematically about misunderstandings. Most teachers are very anxious to help their students learn, and mathematical content—if it is conveyed in that context—offers a way of thinking about teacher preparation and enhancement programs that is potentially of great use. The next agenda for mathematics improvement surely must include these important issues of teacher preparation and professional development. Until the preparation of incoming college students is stronger, which is certainly related to the mathematical power of their teachers, we will never get to the point in undergraduate mathematics education where we would like to be.

This chapter places *Calculus Renewal* in the context of some broader issues in undergraduate SMET education nationally. Out of those broader issues have come several questions that are of great significance for the mathematics community to consider as it decides what will come next after the calculus development of the past decade. It is the hope of this author that the community will address these issues in a sustained and comprehensive way, so that the calculus changes put in place can lead to a stronger and more vital overall undergraduate program in mathematics for *all* students.

REFERENCES

1. National Commission on Excellence in Education, *A Nation at Risk* (U.S. Government Printing Office, Washington, DC, 1983).
2. National Science Board, Task Committee on Undergraduate Science and Engineering Education, *Undergraduate Science, Mathematics and Engineering Education; Role for the National Science Foundation and Recommendations for Action by Other Sectors to Strengthen Collegiate Education and Pursue Excellence in the Next Generation of U.S. Leadership in Science and Technology* (National Science Foundation, Washington, DC, 1986, NSB86-100).
3. National Science Foundation, Advisory Committee for Education and Human Resources, *Shaping the Future: New Expectations for Undergraduate Education in Science, Mathematics, Engineering, and Technology* (National Science Foundation, Washington, DC, 1996, NSF96-139).

4. National Research Council, *From Analysis to Action: Undergraduate Education in Science, Mathematics, Engineering, and Technology* (National Academy Press, Washington, DC, 1996).

5. National Research Council, Mathematical Sciences Education Board, *The Preparation of Teachers of Mathematics: Considerations and Challenges, A Letter Report* (National Academy Press, Washington, DC, 1996).

CHAPTER 2

The Goals of the Calculus Course

William McCallum

It has been more than a decade since the beginning of efforts to renew calculus teaching in this country. Those efforts, successful beyond the wildest dreams of their initiators, have reached a plateau. The reaction has been at times intemperate and at times thoughtful, and has led to serious efforts to reinvigorate the traditional curriculum (see *How to Teach Mathematics*[1] and its appendixes).

The move to change calculus teaching began with a conference held at Tulane University in 1986, funded by the Sloan Foundation,[2] which issued a call for change in the teaching of calculus. This was followed in 1987 by a national colloquium on Calculus for a New Century,[3] sponsored by the National Academy of Sciences and the National Academy of Engineering, and in 1988 by a request for proposals from the National Science Foundation, causing a burst of activity and leading to fundamental changes in curricula and teaching methods across the country.

Why was there a call for change? What were the goals of those who responded to it? What has been achieved? And, finally, where should the mathematical community go now? The purpose of this chapter is to consider these questions and thus initiate a discussion of common goals for calculus teaching.

1. HOW DO WE EXAMINE WHAT WE DO?

A fundamental question we ask ourselves when we look at a mathematics course is: What's in it?

There are two ways to answer this question. First, one can look at the syllabus and the textbook. These tell us what the course intends to cover; however, intentions are not always achieved. A second way to look at a curriculum is to look at the tests (see McCallum[4]). Most of the questions on a test are ones that the instructor expects many of the students to be able to do; thus, the tests give a more accurate idea of what is really in the course. The two ways of looking can give very different pictures; a syllabus full of important and interesting topics can go with exams full of mechanical questions; a syllabus that appears sparse can go with challenging and thought-provoking exam questions.

Both ways of looking are important. The first tells us the *intended curriculum*, the second the *achieved curriculum* (this distinction was made in the Third International Mathematics and Science Study[5]). A well-designed syllabus, or a well-written text, shows the structure of the subject, inspires the teacher with a global vision, and expresses clearly the goals of the course. On the other hand, examining the achieved curriculum keeps us honest and provides us with a reality check. Unfortunately, intentions are usually easier to examine than achievements, so that reality check has sometimes not been present in the arguments about calculus teaching over the last decade.

2. WHERE WE WERE

When I first started teaching calculus in 1979, as a graduate teaching assistant at Harvard University, I was thrilled with the prospect of explaining the concepts of limit, derivative, and integral, and looked forward to the climax of the story, the Fundamental Theorem of Calculus. However, I gradually discovered that teaching calculus was as far removed from my own mathematical experience as teaching neat handwriting would be to a student of English Literature. I discovered that most of my students were attempting to navigate the course entirely by memorizing rules and procedures, and that they were largely succeeding because the homework and exam questions were all cut from a limited set of templates defined by examples in the book.

My experience illustrates the chasm that existed in the calculus course of the 1980s between the intended and achieved curricula. The intended curriculum has many defenders, and rightly so. The story it aims to tell, from the foundations based on limits to the intricacies of sequences and series, is an important and beautiful one. In practice, however, students often learned only the Cliff Notes version of that story. Students were asked to memorize, but not to understand, the theoretical definitions. They were taught a five-step procedure for solving word problems, and then given problems designed for that procedure. They were taught to determine maxima and minima by drawing sign diagrams, but often didn't know that a plus sign meant the graph was sloping up.

Is it possible that I am mistaken about the state of calculus teaching 10–20 years ago? Many defenders of the traditional curriculum think so. They argue that there was much of value in the techniques we taught at that time; that the understanding that reform demands can only come after a solid grounding in calculation skills; that the problem lies elsewhere, in students' work habits and attitude to learning. I respect these arguments, and agree that the mental attitude of students is a large part of our problem. However, I contend that we have been complicit in sustaining that attitude. I remember the prevalent belief that our role was to weed out the weaker among the premeds and budding engineers, rather than to teach our subject.

As for calculation skills, I believe that although we may have done a good job of teaching students to perform certain procedures, we didn't teach them *flexible* skills. Consider, for example, the convergence of an infinite series such as

$$\sum_{n=0}^{\infty} \frac{n!}{(2n+1)(2n-1)\cdots 1}$$

Our best students learned how to use the ratio test to show that it converged. But suppose we vary the problem slightly, and ask for the convergence of $\sum_{n=0}^{\infty} a_n$, where $a_0 = 1$ and $a_n = [n/(2n+1)]a_{n-1}$ for $n > 0$. For a student who understands the ratio test, this formulation should, if anything, be easier than the other; but most students found it harder because it didn't fit their template for a ratio test problem.

Many of the arguments about whether a given topic should be "left in" or "taken out" of the curriculum can be short-circuited by checking to see if the topic was ever in the curriculum at all. For example, two favorite topics of discussion are the theory of limits and the mean value theorem. Looking at the collection in Steen[3] of 25 final exams in Calculus I and II from the late 1980s, we find one question on the former (asking students to use limit laws to evaluate a limit) and two questions on the latter (asking students to "find the c" for a specific function). The theoretical content of even these questions is dubious; there are no questions at all dealing with the ideas behind the formal definition of limit or with the role of the mean value theorem in calculus.

The collection of exams in Steen[3] illustrates another characteristic of the traditional curriculum: its uniformity. This can also be seen in the textbooks of the time, which, as noted in Renz,[6] showed little variability.

3. THE GOALS OF RENEWAL

Those who joined the effort to change the calculus curriculum had a variety of goals, which may be grouped into three broad areas: conceptual understanding, realistic problems, and use of technology.

3.1. Conceptual Understanding

Further, many of those who do finish the [calculus] course, have taken a watered down, cookbook course in which all they learn are recipes, without even being taught what it is that they are cooking. [Ronald G. Douglas, Tulane conference,[2] 1986]

During the late 1980s, some mathematicians became dissatisfied with teaching a course that had so little mathematics in it. The first step in restoring understanding was to make sure that students had a good intuitive grasp of the concepts, and that they could provide good heuristic arguments for basic theorems such as the fundamental theorem of calculus. Many of the calculus reform projects put an emphasis on using graphical and numerical work to provide the intuition and heuristic arguments.

Let me give an example of this: determining the convergence or divergence of the improper integrals

$$\int_1^\infty \frac{1}{x}\,dx \quad \text{and} \quad \int_1^\infty \frac{1}{x^2}\,dx$$

By graphing the integrands successively on the intervals $[10^n, 10^{n+1}]$, $n = 0, 1, 2, \ldots$, one sees that the first requires scaling the units on the vertical axis down by a factor of $1/10$ each time in order for the graph to remain visible above the x-axis, whereas the second requires a scaling of $1/100$. Since the x-range of each graph increases by a factor of 10 each time, one sees why the areas under the first graph grow without bound, whereas those under the second compare with a geometric series with common ratio $1/10$. This provides a concrete understanding of the convergence properties, which both complements and is confirmed by the algebraic calculation of the integrals.

In addition to giving scientists and engineers the understanding to make intelligent use of the tools of calculus, an approach based on intuitive geometrical and numerical reasoning makes a better foundation for rigorous mathematics than superficial exposure to formal definitions and proofs, as I have argued elsewhere.[7]

An important feature of the new curricula is that they require students to give coherent written explanations of their reasoning, using their intuitive under-standing (this is a particular feature of Duke University's Project CALC[8]). As a result, students often find the new courses more challenging than the traditional course, since the homework and test questions are more demanding. However, because formal definitions and proofs are omitted, many mathematicians who look only at the topics in the table of contents wrongly conclude that the new curricula constitute a dilution of the traditional one.

3.2. Realistic Problems

> Whereas mathematics has most often been considered as a requirement for *knowledge*, it is time for us to begin to consider its role in *judgment*. [W. Dale Compton, NAS–NAE colloquium,[3] 1987]

W. Dale Compton is a Senior Fellow at the National Academy of Engineering. What is striking to the mathematician reading his statement above is the implicit charge that judgment had not hitherto been playing a role in mathematics courses. But, indeed, students entering science and engineering courses were often unable to use mathematics outside a narrow range of artificially constructed problems; it is a natural assumption on the part of engineers and scientists that this inability reflected the training we mathematicians had given them. Many in the reform movement sought out people in other disciplines and asked how they used mathematics and what their students had trouble with. Some of the changes that resulted from this dialogue were simple: Don't always call the variables x, t, or y; put more emphasis on units. Others were more fundamental: Teach differential equations earlier; include examples from biology and economics.

Concomitant with the move to more realistic applications was a move to more extended problems or projects, requiring many steps and sustained consideration over a number of days, rather than problems that could be solved in one sitting by studying the method of an example in the book. Since many heads are better than one in approaching challenging problems, a natural consequence was experimentation with group work.

Realistic problems also reinforced the demand for more writing from the students. Whereas a chicken-scratch of equations was formerly accepted as an answer, students were now required to explain the meaning of their answers in practical terms. This helped develop purely mathematical understanding, as well as practical scientific understanding.

3.3. Use of Technology

> The single most striking new development in [mathematics and the sciences] is the rise, indeed ubiquity, of computers of high speed and large capacity. They make possible numerical and symbolic explorations on an unprecedented scale [Peter Lax, Tulane conference,[2] 1986]

Most of those interested in changing the curriculum grappled with the challenge presented by computer technology in one way or another. For example, Paul Zorn became interested in using symbolic manipulators to shift the emphasis away from techniques and toward concepts; Thomas Dick sought to use the graphical capabilities of handheld calculators to support geometric and intuitive understanding. Jerry Uhl took to heart Peter Lax's call for a more modern curriculum incorporating computing, and later discovered the potential for using computers to aid in the "visual acquisition of ideas." Almost all reformed calculus curricula made some use of the graphical and numerical capabilities of technology, which have now spread to traditional texts as well. However, most have not yet grappled with the issue of symbolic manipulators (*Calculus & Mathematica*[9] is an exception).

4. WHERE WE ARE NOW

The calculus reform movement changed the way many people teach calculus. As a result, there is now more than one choice of calculus course. There has been a tremendous amount of discussion and argument about how to teach calculus: Should we use technology or not? Should we emphasize mechanical skills or conceptual understanding? What should be the balance between applications and theory? Between formal and informal arguments? Between the traditional lecture method and more interactive approaches? I was brought up in a large and noisy family, where survival at the dinner table depended on one's ability to defend any position, whether one agreed with it or not. I have a fateful ability to see both sides of almost all of the arguments about calculus teaching. However, what's different now from 10 years ago is that on almost every one of these questions, there is now a spectrum on which one can place oneself; there is a much wider acceptable range of opinions.

Another achievement of the calculus reform movement is that people are now talking about how to teach in a way that was unheard of a decade ago. Hyman Bass of Columbia University put it this way:

> What [the people engaged in calculus reform] find most significant about the reform is their personal transformation and the change in their professional practice as teachers. They gain a sense of having become members of a community for which the practice of teaching has become a part of professional consciousness and collegial communication, not unlike their professional practice of mathematics itself. It is the creation of this substantial community of professional mathematician-educators that is, to my mind, the most significant (and perhaps least anticipated) product of the calculus reform movement. This is an achievement of which our community can be justly proud, and which deserves to be nurtured and enhanced.[10]

5. WHERE ARE WE HEADED?

As we turn the century, two big challenges face mathematics educators, traditionalists or reformers. The first is the problem of standards.

Although traditional texts have made some moves in the direction of reform, for example by adding more graphical and numerical work, they still retain a largely template-driven approach to the exercises. Furthermore, calculus reform texts will face mounting pressure to make the life of students and faculty easier by providing more examples, more standard problems, more ways for students to navigate the course without thinking and for faculty to teach it without too much preparation or too many office hours.

The second challenge is technology. As cheap symbolic manipulators become more readily available, we must think about their appropriate use. Banishing them from the classroom will leave students to learn their use without our guidance; accepting them into the classroom requires decisions about which symbolic skills are purely a matter of computation and can safely be left to the computer, and which are the basis for conceptual understanding.

What, then, are the goals of the calculus course? It is neither possible nor desirable to give a definitive listing of topics; indeed, the wide range of choices now available should be preserved. However, it is necessary to frame some basic principles against which to measure our continued efforts to renew calculus teaching.

Goals of the Calculus Course

At the end of a successful first-year calculus course, students should be able to:
1. Make calculations with agility, accuracy, intelligence, and flexibility.
2. Explain the basic concepts of calculus clearly and reason mathematically with them.
3. Solve extended problems, with good judgment in the choice of tools and in checking answers.
4. Make connections between different incarnations of the same idea.
5. Use calculus to model realistic situations from engineering and from the physical, life, and social sciences.

- *Make calculations with agility, accuracy, intelligence, and flexibility.*

These calculations include solving equations, and finding derivatives, antiderivatives, and definite integrals. Traditionally, calculations were made by pencil and paper, using algorithms instilled by drill; nowadays, computers provide an alternative method of calculating. There has been much debate about which methods of calculation students should be permitted or required to use, and when. It would be more fruitful to focus less on which methods are used, and more on how they are used. I propose four criteria for evaluating competing claims about calculation methods: agility, accuracy, intelligence, and flexibility. The first two are clear enough; students should be able to get it quickly, and get it right. Intelligent calculation means being able to select the right method and knowing how to check the answer—the opposite of both mindless button pushing and mindless pencil pushing. Flexibility implies the ability to adapt computational methods to different situations; for example, understanding substitution as a general method, in addition to simply knowing a lot of substitutions.

These four criteria do not particularly favor one method over another. For example, a student who evaluates $\int \cos(10x)dx$ by hand is likely to be quicker and more accurate than one who uses a machine. Furthermore, such a student is in a better position to analyze the behavior of $\int_0^{\pi/2} \cos(ax)dx$ as a grows large. On the other hand, a student who integrates $\int \cos^{10}(x)dx$ using the recursion formula is likely to be slower and less accurate than one who uses a machine. Furthermore, the latter student is probably in a better position to conjecture the general shape of $\int \cos^n x \, dx$.

- *Explain the basic concepts of calculus clearly and reason mathematically with them.*

Although there is widespread agreement that conceptual understanding is an admirable intention of the calculus course, there is less agreement about putting it on the final exam, so to speak. I have often heard the argument that it is unreasonable to expect students to understand everything the first time, and that the best we can do is make sure that they can follow procedures the first time around, leaving understanding to come later. This echoes the "spiral curriculum" approach to K–12 education, which advocates repeated visits to the same topic, at an incrementally higher level each time. As the TIMSS study has shown, U.S. schools leave material in the curriculum longer than most countries, without good results.[5]

Although some concepts will not be grasped by all students the first time they are encountered, it nonetheless is important to make a demand for understanding. That demand may not always be met, but it will prevent the spiral from collapsing to a circle. Therefore, I contend that every concept we regard as important has to be incorporated into the course at a level beyond mere exposition. If it's worth teaching, it's worth testing on an exam. For example, students should not merely know the fact that an inflection point occurs where the second derivative changes sign, but be able to explain why this is so in terms of increase and decrease of the first derivative.

- *Solve extended problems, with good judgment in the choice of tools and in checking answers.*

Once students have acquired the ability to make calculations, and have grasped the basic concepts, they should be able to make use of their ability and understanding in solving extended problems. By an extended problem, I mean one that requires the application of a sequence of techniques and concepts, and possibly admits more than one such sequence. It is important that students be able to make their way through such problems without too much guidance. Too often exam questions are sprinkled with the word "use": "Use the fundamental theorem of calculus to...," "Using a Riemann sum...." Such hints have the effect of breaking the problem down into manageable one-step exercises, much as a driving instructor uses independent controls to keep a beginning student from crashing. Although this is good for beginners, students should finally make the transition from following directions to making their own way.

As any traveler knows, the most stressful points in a journey are often the departure and the arrival. At the beginning, many difficult choices must be made; at the end, we must face up to the consequences of those choices, and be willing to backtrack if necessary. Students should learn to make their own journeys through mathematical problems, by making independent choice of the equipment they will need and by learning to check their own answers.

- *Make connections between different incarnations of the same idea.*

The value of multiple representations lies as much in the effort required to make connections between them as in the intrinsic value of the representations themselves. This point is well illustrated by the following example, which I owe to Barbara Shipman. As a graduate teaching assistant at the University of Arizona, Barbara once asked her class to plot the parametric curve $x = \cos(t^3)$, $y = \sin(t^3)$, $0 \le t \le 10$, on their graphing calculators. The students obtained something like the picture in Fig. 1.

This led to a discussion of what the graph should look like; eventually the students concluded, with no help from Barbara, that the identity $\sin^2(x) + \cos^2(x) = 1$ meant the graph had to lie on a circle, no matter what the calculator showed. Finally, taking a numerical look at the change in position of the point (x, y) with increments in t led to an understanding of the strange picture on the calculator.

The goal of making connections is also important in courses operating at a high level of mathematical rigor. Much of the beauty of formal definitions and theorems lies in the simple way they capture in logical language our geometric and numerical understanding. Thus, for example, in considering the $\epsilon - \delta$ definition of limit, it is not only important that the student be able to use the definition correctly in proving statements about limits, but also that the student

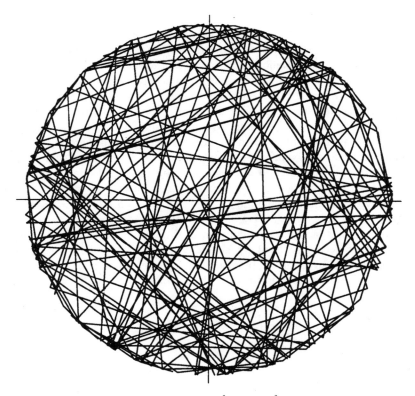

Figure 1. Graph of $x = \cos(t^3)$, $y = \sin(t^3)$, $0 \leq t \leq 10$.

understand the definition as a reincarnation of the more intuitive geometric conception. Without that ability to make the connection to the intuitive point of view, there is little chance of moving beyond simple formal manipulations of the definitions.

- *Use calculus to model realistic situations from engineering and from the physical, life, and social sciences.*

Many of our students take calculus for the purpose of using it to model scientific or engineering problems. The classical application of calculus to such fundamental problems as planetary motion is an important part of its historical development, and modern applications to fluid dynamics or population biology, for example, inspire students with the power of calculus as an analytical tool. Even for students going on to theoretical mathematics, it is important to understand the roots of calculus in science; furthermore, the mental exercise of thinking through the interpretation of a concept in realistic contexts helps build understanding of the concept itself. For example, understanding the density interpretation of the space derivative of a mass function throws light on the definition of the derivative.

6. CONCLUSIONS

The goals proposed here are sufficiently broad to allow a wide variety of courses, but sufficiently tight to ensure decent standards for student ability and knowledge. They do not explicitly advocate any particular method of teaching, although some methods may turn out to be better than others for achieving them. They do require three broad areas of content—technical skills, concepts, and applications—without prescribing specific topics within those areas. Most importantly, the goals specify a certain intellectual level for the calculus course; students should understand what they are doing, and be able to use what they have learned independently of the instructor and in unfamiliar contexts.

Setting goals is one thing; finding practical steps to achieve them is another. I would like to conclude this chapter by returning to a point made at the beginning, which suggests one such step. Two different ways of looking at the curriculum were discussed:

- Textbook and syllabus
- Homework problems and exams

If there is one single change in practice that would move us closer to the goals proposed in this chapter, it is to systematically use the second way of looking in our deliberations (not, of course, to the exclusion of the first). This means that when we consider adding a new topic to a course, we should ask ourselves: What problems will I give the students about this topic? It means that when we go to other disciplines for their views on the calculus course, we should ask for problems they would like their students to be able to do, in addition to asking for lists of topics. It means that we should pay more attention to the design and variation of problems; that we should not lead students to expect that a problem will always fit into a previously learned pattern, or that there will always be a hint suggesting the method to be used. It means that we should use problems as a tool for developing intuitive understanding, as well as a method for testing skills. It means that in evaluating different approaches to teaching, we should look not only at the table of contents of a text or the brand of technology used in a lab, but at the homework exercises in the text or the way students use the technology. Finally, it means paying more attention to the answers our students write. We should treat those answers seriously by expecting them to be coherent, self-contained, written in clear language, and demonstrating an awareness of the meaning of calculation results. We should grade those answers seriously by not filling in the gaps ourselves and providing partial credit for scattered fragments of correct calculation.

A course meeting the goals described in this chapter is *worth teaching* and *worth taking*; it requires that teachers bring to it a thorough understanding of how calculus works and how students think; and it requires that students bring to it a willingness to work hard at a deep intellectual endeavor.

REFERENCES

1. S. G. Krantz, *How to Teach Mathematics*, 2nd ed. (American Mathematical Society, Providence, 1999).
2. R. G. Douglas (ed.), *Toward a Lean and Lively Calculus*, MAA Notes #6 (Mathematical Association of America, Washington, DC, 1986).
3. L. A. Steen (ed.), *Calculus for a New Century: A Pump, not a Filter*, MAA Notes #8 (Mathematical Association of America, Washington, DC, 1988).
4. W. G. McCallum, "Will This Be on the Exam?" In Ref. 1.
5. Third International Mathematics and Science Study, http://nced.ed.gov/TIMSS.
6. P. Renz, "Style Versus Content: Forces Shaping the Evolution of Textbooks," pp. 85–102 in Ref. 2.
7. W. G. McCallum, "Rigor in the Undergraduate Calculus Curriculum," *Notices of the American Mathematical Society* 38(9)(1991):1131–1132.
8. D. Smith and L. Moore, *Calculus: Modeling and Application* (Heath, Boston, 1996).
9. J. Uhl, W. Davis, and H. Porta, *Calculus & Mathematica*: (Addison–Wesley, Reading, MA, 1994).
10. H. Bass, "Mathematicians as Educators," *Notices of the American Mathematical Society* 44(1)(1997):18–21.

CHAPTER 3

Renewal in Collegiate Mathematics Education

Learning from Research*

David A. Smith

To the memory of James R. C. Leitzel

1. CALCULUS: REFORM OR RENEWAL?

Pop quiz: Who wrote this and when?

> It is to be hoped that the near future will bring reforms in the mathematical teaching in this country. We are in sad need of them. From nearly all of our colleges and universities comes the loud complaint of inefficient preparation on the part of students applying for admission; from the high schools comes the same doleful cry. Educators

*This is a somewhat expanded version of a paper with a similar title first published in *Documenta Mathematica*, Extra Volume ICM 1998 III, pp. 777–786. The previously published version is available at http://www.math.uiuc.edu/documenta/xvol-icm/18/Smith.MAN.html. I thank the many colleagues who have written thoughtful commentaries on that version, especially Mark Bridger, Murray Eisenberg, Cathy Kessel, Michael Livshits, Richard Mercer, David Olson, Susan Pustejovsky, Nell Rayburn, Lynne Small, Lynn Steen, and Tina Straley. If their suggestions are not yet fully implemented, that is not because I reject their opinions but rather because I still have a lot to learn about the subject treated here.

who have studied the work of [foreign] schools declare that our results in elementary
instruction are far inferior.

The answer[1] illustrates that questions of reform and renewal in mathematics
education are not new—this effort has continued with varying levels of intensity
for more than a century.

"The great obstacle to progress is not ignorance but the illusion of knowledge."*

We take for granted that anyone with a Ph.D. in mathematics has mastered
the subject and is prepared to teach at the university level. After all, we had
excellent role models. If we do what they did, we will be successful—it worked
for us. This is not ignorance but a dangerous illusion of knowledge: Good
teaching engendered learning in us, so our job is good teaching—learning will
follow.† If it doesn't, the students must be at fault.

In the mid-1980s there was widespread recognition that something was
wrong with this theory, at all levels of mathematics education. Calculus was
chosen as the primary target for "reform" in part because it was both the capstone
course for secondary education and the entry course for collegiate mathematics.
Thus was born the calculus reform movement, whose history, philosophy, and
practice are described in Solow,[4] Tucker and Leitzel,[5] and Roberts.[6]

The first National Science Foundation calculus grants were awarded in 1988.
Since then we have seen development and implementation of several new
approaches to teaching calculus, with widespread acceptance on some campuses,
and rejection and backlash on others. Our own approach (Project CALC) is to
treat calculus as a laboratory science course that emphasizes real-world problems,
hands-on activities, discovery learning, writing, teamwork, intelligent use of
tools, and high expectations of students.

In hindsight, "reform" was not a good choice of name. The word has stuck,
and most people recognize the course types to which it currently refers. However,
it is an emotionally charged word—religious wars have been fought over it. One
source of the current controversy is that people with deeply held beliefs feel they
are under attack. "Renewal" is a better descriptor—perhaps we can discuss
rationally whether the new aspects of mathematics teaching are also *good*, and
whether renewal of pedagogical strategies from time to time is itself a good thing
to do.

At the time of development, the Project CALC team had little or no
theoretical support for our choice of strategies. In place of theory, we relied on
careful empirical work. The main body of this essay suggests a possible
development of the theoretical base that we lacked 10 years ago. Results from

* Daniel Boorstin, September 9, 1987, in a television interview on the occasion of his retirement as
director of the Library of Congress (quoted in Chickering and Gamson[2], p. 57).
† For a counterexample, see Schoenfeld.[3]

cognitive psychology were in the literature then but unknown to us and most of the other developers. Results from neurobiology have come to fruition just in this decade, and they tend to confirm cognitive theories that fit with our empirical observations. We are in the process of replacing the illusion of knowledge with real knowledge about learning and the teaching strategies that engender learning. But this process is still at an early stage—psychologists and biologists argue among themselves and with each other about proper interpretations of their research, and there is certainly no consensus about application of this research to education. Thus, one should take the nonexpert assertions in this essay as a plausible scenario—possibly leading to a research agenda—for bringing together significant ideas from psychology, biology, and education.

2. WHO STUDIES CALCULUS AND WHY

Some 700,000 students are enrolled in college-level calculus courses in the United States in any given year. Of these, 100,000 are in Advanced Placement courses in high schools, 125,000 in two-year colleges, and the rest in four-year colleges or universities.[4] A very small percentage of these students intend to take any mathematics beyond calculus, let alone major in mathematics or do graduate study or become a mathematician. Most of this enrollment is generated either by general education requirements or by prerequisites for subsequent course work. To cite just one example, Duke University has flexible distribution requirements that allow students to avoid calculus, but it also has 24 major programs that require one or more semesters of calculus. Even though many students enter with Advanced Placement credits, some 80% of our first-year students take a calculus course. About 2% of each class graduates with a major in mathematics. Thus, most students are not motivated to study calculus except as it serves some other goal, e.g., keeping open options for a major.

U.S. colleges provide liberal, vocational, and/or preprofessional education to students who overwhelmingly see themselves as participants in preprofessional or vocational programs. A small percentage contemplate academic graduate study, but only the tiniest fraction have any concept of liberal education and its potential importance in their lives. Parents, who pay the bills through tuition and/or taxes, usually see things the same way: The objective is for their child to become productive and self-supporting. And potential employers of graduates at all levels have definite expectations for the skills and abilities of their employees. Since these employers collectively influence support for and accountability from institutions of higher education, public or private, it behooves us to pay attention to what they want.

Here is what employers want,[7] expressed in seven "skill groups":

1. The foundation: knowing how to learn
2. Competence: reading, writing, and computation
3. Communication: listening and speaking
4. Adaptability: creative thinking and problem solving
5. Personal management: self-esteem, goal setting and motivation, personal and career development
6. Group effectiveness: interpersonal skills, negotiation, and teamwork
7. Influence: organizational effectiveness and leadership

With the possible exception of item 2, this is very different from what educators usually think of as "basic skills." If students enter college lacking most of these skills, where are they supposed to learn them? In college, obviously. Indeed, this list defines the goals of higher education in the broad sense: liberal, vocational, and preprofessional. Whose job is it to teach these skills? Clearly it is the job of the entire faculty, including the Mathematics Department—and not just for "computation" and "problem solving." To get a consistent message from the faculty and to have a good chance of graduating with these skills in place, students must encounter most of them in almost every course.

Some institutions, such as Alverno College[8] and Evergreen State College,[9] have clear expectations that their courses will address a similar list of learning objectives, so we know this is possible. Professional schools are more directly accountable for these outcomes than are liberal arts schools, so it is no accident that many of them are ahead of the undergraduate institutions in transforming themselves to achieve these objectives. The story of such a transformation at the management school of Case Western Reserve University is told in Boyatzis *et al.*[10] We will see in later sections why addressing these objectives enables more students to learn more mathematics.

In addition to the needs of employers, who may be seen as "end users" of our "products," we must pay attention to the needs of "intermediate users": the academic disciplines that require their majors to study calculus. These needs are often cited as reasons why no significant change is possible—as in, "I would like to use [substitute your favorite change], but our engineers would never stand for it." It's true that colleagues in other departments (as in our own) sometimes act more like anchors than like sails. But they may be out of touch with their own disciplines as well as with ours. All accredited engineering programs are subject to the ABET requirements,[11] a new set of which is currently being phased in. These new standards require accredited programs to demonstrate that their graduates have, among other things, abilities to

- Apply knowledge of mathematics, science, and engineering
- Analyze and interpret data
- Function on multidisciplinary teams

- Identify, formulate, and solve problems
- Communicate effectively
- Engage in lifelong learning

These new standards are largely concordant with the employers' seven skill groups—but much more specific to the needs of engineering—and the items listed here are addressed explicitly in many reformed mathematics courses. Thus, renewal efforts in mathematics and in engineering are moving in the same direction and serve to reinforce each other. Unlike mathematics departments, however, engineering departments have no choice about whether to make these changes.

3. PROBLEMS WITH COLLEGIATE EDUCATION IN MATHEMATICS

What was wrong with mathematics education in colleges and universities in the 1980s that led to a perceived need for reform? Many have described the turned-off students and jaded faculty in our classrooms and lecture halls, usually with the intention of blaming someone—teachers at a lower level, society, administrators, or the students themselves. A more constructive description appears in a recent essay from the Pew Higher Education Roundtable,[12] a product of discussions among a group of 35 science faculty, administrators, foundation officers, and program directors. Their thesis is that there is broad consensus on what constitutes effective science education, but institutional barriers to change have thus far prevented widespread implementation. In the following quote from that discussion, the word *science* is shorthand for "science, mathematics, engineering, and technology," all of which have had similar problems with education—with similar solutions.

> The traditional approach is to conceive of science education as a process that sifts from the masses of students a select few deemed suitable for the rigors of scientific inquiry. It is a process that resembles what most science faculty remember from their own experiences, beginning with the early identification of gifted students before high school, continuing with the acceleration of those students during grades 9 to 12, fostering in them the disciplined habits of inquiry through their undergraduate majors, and culminating in graduate study and the earning of a Ph.D. Forgotten are . . . most students for whom a basic knowledge of science is principally a tool for citizenship, for personal enlightenment, for introducing one's own children to science, and for fulfilling employment. Forgotten as well are those students who will become primary and secondary school teachers and, as such, will be responsible for the general quality of the science learning most students bring with them to their undergraduate studies. . . .
>
> Although it is widely recognized that an inquiry-based approach to science increases the quality of learning, introductory-level students are often not given to understand what it means to be a scientist at work. . . .

... science faculty have at times openly acknowledged their tendency to gear instruction to the top 20 percent of the class—to those students whose native ability and persistence enable them to keep pace with the professor's expectations. The fact that others are falling behind and then dropping out is seen not as a failure of pedagogy but as an upholding of standards.

In short, when we use ourselves as models for our students, we get it all wrong. Hardly any entry-level mathematics and science students are like us. In particular, most students in most calculus courses are in their *last* mathematics course. And these students are not *our* replacements. Rather, they are the next generation's parents, workers, employers, doctors, lawyers, schoolteachers, and legislators. So it matters to our profession how they regard mathematics.

It's not hard to trace how we got to be so out of touch with the needs of our students. Those of us educated in the Sputnik era were in the target population of that "traditional approach" described above—just at the end of a time when it didn't matter much that the majority of the college-educated (an elite subset of the whole population) didn't know much about science or mathematics. As we became the next generation of college faculty, the demographics of college-bound students broadened significantly, new money flowed to support science more generously, and broad understanding of science became much more important. The reward structure for faculty was significantly altered in the direction of research—away from teaching—just when we were confronted with masses of students whose sociology was quite different from our own.

This oversimplifies a complex story, but the response of the profession was to water down expectations of student performance, while continuing to teach in the only way we knew how—as we were taught. We created "second-tier" courses (e.g., calculus for business and life sciences), we wrote books that students were not expected to read, and even in our "mainline" courses there were questions we didn't dare ask on our tests. The goal for junior faculty was to become senior faculty so we wouldn't have to deal with freshman courses. Along the way, we produced high-quality research and excellent research-oriented graduate students to follow in our footsteps. But seldom in our graduate or professional careers was there any opportunity or incentive to learn anything about learning—in particular, about how our students would be approaching that task.

The problems identified by the Pew Roundtable are not unique to science and mathematics, but rather fit neatly into the context of higher education problems in general. Diamond[13,14] notes that there is actually broad consensus among faculty on the goals of higher education—and among students, parents, employers, legislators, and academics that far too many students graduate without mastering the core skills. The consensus goals include skills in

- Communication (writing, speaking, reading, listening)

- Mathematics (especially basic statistics), problem solving, critical thinking
- Interpersonal relations (e.g., working in and leading groups)
- Computer literacy
- Appreciation of diversity and adaptation to innovation and change
- Knowledge in the major discipline
- Some general knowledge in other core disciplines

Diamond's reasons why we are unable to achieve these goals are these:

- Higher education rarely deals with the goals of instruction directly and has avoided stating them in measurable terms.
- Courses and programs rarely are designed around providing each student the means to attain the competencies we agree on.
- Faculty receive little reward for devoting significant amounts of time and energy to improving courses and curricula.
- Higher education has been able to avoid hard questions about assessment of educational outcomes—and so it has avoided them.
- Many who might like to design better courses and curricula don't know how to do so very well.

This list includes major issues of both politics (e.g., rewards and accountability) and substance (e.g., design and assessment). The remainder of this essay will focus on the latter set of issues.

4. SOME MESSAGES FROM COGNITIVE PSYCHOLOGY

Starting in 1985, Chickering and Gamson[2] led an extensive effort, with an exhaustive review of "50 years of research on the way teachers teach and students learn" and a conference of leading educators and researchers, to define what works in college education. The upshot was the publication of the following Seven Principles and supporting documentation (Appendix A in Chickering and Gamson[2]):

Good practice in undergraduate education:

1. Encourages student–faculty contact
2. Encourages cooperation among students
3. Encourages active learning
4. Gives prompt feedback
5. Emphasizes time on task
6. Communicates high expectations
7. Respects diverse talents and ways of learning

Not long after, they published two detailed inventories—for faculty and institutional administrators (Chickering and Gamson,[2] Appendices B and C)—that could be used to assess the extent to which a school, its departments, and its faculty do or do not follow these principles. One does not need an inventory to see that much of the traditional teaching practice in mathematics is not in accord with these principles. But it doesn't have to be that way. Indeed, Chickering and Gamson[2] is a straightforward handbook for implementing these principles.

Research in cognitive psychology has been sending consistent messages for a half-century or more, but it is clear that very few mathematicians have been listening until the current decade. As Chickering and Gamson summarize,

> While each practice can stand on its own, when they are all present, their effects multiply. Together, they employ six powerful forces in education:
>
> - Activity
> - Cooperation
> - Diversity
> - Expectations
> - Interaction
> - Responsibility.

Cognitive psychology research contains many competing models of the learning process. There is not room here to compare and contrast these, but we describe one, the Kolb learning cycle (pp. 128–133 in Kolb et al.[15]), as a way to think about our students and our relationship to them (see Fig. 1). The four stages of this cycle are

- Concrete Experience (CE)
- Reflection/Observation (RO)
- Abstract Conceptualization (AC)
- Active Experimentation (AE)

The ideal learner cycles through these stages in each significant learning experience. The AE stage represents testing in new situations the implications of concepts formed at the AC stage. Depending on the results of that testing, the cycle starts over with a new learning experience or with a revision of the current one. The ideal learning environment is designed to lead the learner through these stages and not allow "settling" in a preferred stage.

In fact, there are very few ideal learners. Most of us (and our students) have preferred learning activities and styles, and they are not all alike. This is one reason why learning experiences work better for everyone in a diverse, cooperative, interactive group. (The preceding sentence mentioned three of Chickering and Gamson's "powerful forces," and the other three are there implicitly.)

In the plane of the Kolb cycle (Fig. 1) there are two primary axes, the action–reflection axis (AE–RO) and the concrete–abstract axis (CE–AC). Those axes

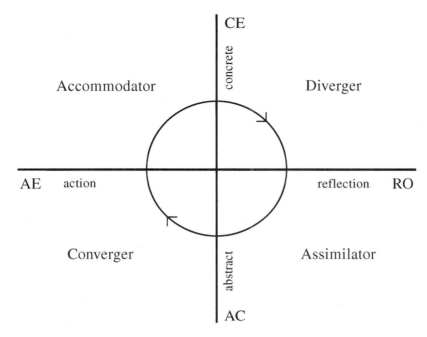

Figure 1. Kolb learning cycle.

divide the cycle into four quadrants associated with four dominant learning styles (Kolb *et al.*,[15] pp. 131–132):

- *Converger* (AC, AE): strengths include practical applications, single correct answers; unemotional, prefers things rather than people; deductive, focuses on specific problems; likely to specialize in physical sciences or engineering.
- *Diverger* (CE, RO): strengths include imagination, organization of many perspectives into gestalt; emotional, prefers people rather than things or theories; imaginative; likely to specialize in arts or humanities.
- *Assimilator* (AC, RO): creates theoretical models; prefers abstract concepts rather than applications or people; inductive, assimilates observations into integrated explanations; likely to specialize in basic sciences or mathematics.
- *Accommodator* (CE, AE): strengths include doing things, carrying out plans; impatient; prefers taking risks; intuitive, operates on trial and error; likely to specialize in business.

Most people are not rooted at a single point in the learning style plane, but rather move around in some subset of this plane, depending on the task at hand. However, it is safe to say that most mathematicians spend much of their time in the Assimilator quadrant, whereas the students in a calculus class are likely to come from at least three and maybe all four quadrants. If our pedagogical strategies address only the students who are "like us," we are not likely to succeed in reaching all of the students who expect to learn from us.

5. SOME MESSAGES FROM MODERN BRAIN RESEARCH

This is the Decade of the Brain, an exciting period of advances in neurobiology that (along with work in related areas, such as genetics) may lead to biology dominating the scientific agenda of the twenty-first century in the same way that physics dominated the twentieth. This work builds on research with animal models and with epileptics who have had split-brain surgery, but the most exciting advances have come from imaging techniques—CAT, PET, MRI. Scientists can now study functioning human brains, normal and otherwise. For example, this research has shown that serious mental disorders, such as schizophrenia and bipolar disorder, are biological in origin, and this has led to new classes of psychoactive drugs that are much more effective at controlling symptoms of these disorders. More to the point at hand, we now have some biological insights into the processes of reasoning, memory, and learning in the normal brain.[16]

An important message of brain research for learning is "selection, not instruction." To explain this, Gazzaniga[17] uses an analogy with immunology. It used to be thought that challenges to the immune system "instructed" it to create antibodies. But we now know that we have at birth all of the antibodies we will ever have, and the external challenges select certain antibodies to become active, much as environmental pressures result in natural selection of some traits and suppression of others in evolution. Indeed, evolutionary theory tells us that at birth we have not only our entire immune system, but our entire neural system as well—and that neither has changed significantly in the last 10,000 years.

Learning takes place by construction of neural networks. External challenges (sensory inputs) select certain neural connections to become active, and this is a random selection among many possible connections that could occur, not something that happens by deterministic design. The sensory input enters the brain through old networks—there aren't any others. The input can trigger either memory, if it is not new, or learning, if it is new. The cognitive psychology term for this process is *constructivism*: The learner builds his or her own knowledge on what is already known, but only in response to a challenge or "disequilibration."

In particular, knowledge is not a commodity that can be transferred from knower to learner.

Selection also means that some potential neural pathways are *not* selected, that is, they become dormant through lack of use. Infants have the neural equipment to learn *every* human language, and they do learn the one(s) that occur in their environment. But they don't learn their "native" language by instruction or imitation. (Proof: Babies don't speak the "baby talk" they hear from their parents.) Their abilities to learn unheard languages atrophy, and, as we know, the later one tries, the harder it is to learn another language. The message for collegiate education: If we want to foster such skills as problem solving, creative thinking, and critical thinking, our task is much easier if educational challenges have been developing these skills from infancy. We have a stake in what happens at all levels before college.

Memory is an intricate collection of neural networks. Most experiences initially form relatively weak neural connections in "working memory," which is necessarily of short duration. The biochemical connections become stronger with use, weaker with disuse. The stabilized networks of long-term memory are accessed mainly by numerous connections to the emotional centers of the brain, but working memory has hardly any connections to the emotional brain. That is, working memory is not related to emotions—just facts—but formation of long-term memory strongly involves emotion.[18] The message: We need to stimulate emotional connections to our subject matter if we expect it to transfer to long-term memory.

Similarly, there are strong connections between the emotional and rational centers in the brain. Indeed, emotional pathways can sometimes direct rational decision making before the learner is consciously aware of the decision process. It's not hard to see the evolutionary connection here. Since all of these structures are 10,000 years old, they are intimately related to fight-or-flight reactions and other survival strategies.

Just as emotion is linked in the brain to learning, memory, and rationality, so are the motor centers of the brain, and by extension, the rest of the body. Body movement facilitates learning. Sitting still inhibits learning.[19]

We have already linked brain research to constructivism. Now we connect with Kolb's learning cycle. The CE phase is input to the sensory cortex of the brain: hearing, seeing, touching, body movement. The RO phase is internal, mainly right-brain, producing context and relationship, which we need for understanding. Because the right brain is slower than the left, this takes time. The AC phase is left-brain activity, developing interpretations of our experiences and reflections. These are action plans, explanations to be tested. They place memories and reflections in logical patterns, and they trigger use of language. Finally, the AE phase calls for external action, for use of the motor brain. Thus, deep learning, learning based on understanding, is *whole brain* activity. Effective teaching must involve stimulation of all aspects of the learning cycle.*

6. TECHNOLOGY AND LEARNING

In the minds of many, "reform" in calculus instruction is strongly associated with introduction of electronic technologies: graphing calculators, symbolic computer systems, the Internet. It is a historical fact that these technologies have become widely available, increasingly powerful, and increasingly affordable during the same decade as reform efforts. Is this good or bad or neutral for education?

The short answer is "yes"—that is, use of technology is good or bad or neutral, depending on who's doing what. There is already an embarrassingly large literature addressing such questions as "Do students learn better with calculators (or Maple, or whatever)?," questions that are just as meaningless as they would have been for earlier technologies, such as blackboards, pencil and paper, slide rules, textbook graphics, or overhead projectors. There are also substantial numbers of thoughtful papers that compare particular classroom technology experiments with traditionally taught classes and measure whatever can be measured. The typical conclusion is that students in the experimental group did as well (or only slightly worse) on traditional skills, and they learned other things as well.

There are, of course, costs associated with new technologies, just as there were with older technologies that we now take for granted. There are hardly any believable studies of cost-effectiveness of new (or old) technologies—not because we don't know the costs or won't admit to them, but because we don't have good ways to measure *effectiveness* of education. Think again about the goals in Section 2 (or any alternate set of goals). Our effectiveness at addressing those goals may not be known until long after the students have left us, and maybe not even then.

A more productive line of inquiry is to examine the costs of *not* using technology, in light of the current context of education, of reasonable projections about the world our students will live in, and of what we now know about learning.

Technology is a fact of life for our students—before, during, and after college. Most students entering college now have experience with a graphing calculator, because calculators are permitted or required on major standardized

---

* In addition to the references already mentioned, Damasio[20] and Edelman[21] are useful for understanding brain research and its likely relevance to learning. Sylwester[22] is a non-technical (but well-
   documented) presentation for educators of the potential implications of this research for curriculum
   and pedagogy.

tests. A large and growing percentage of students have computer experience as well—at home, in the classroom, or in a school or public library. "Surfin' the 'Net" is a way of life—whether for good reasons or bad. Many colleges require computer purchase or incorporate it into their tuition structure. Where the computer itself is not required, the student is likely to find use of technology required in a variety of courses. After graduation, it is virtually certain that, whatever the job is, there will be a computer close at hand. And there is no sign that increase in power or decrease in cost will slow down any time in the near future. We know these tools can be used stupidly or intelligently, and intelligent choices often involve knowledge of mathematics, so this technological environment is our business. Since most of our traditional curriculum was assembled in a precomputer age, we have a responsibility to rethink whether this curriculum still addresses the right issue in the right ways—and that is exactly what has motivated some reformers.

But calculus renewal is not primarily about whether we have been teaching the "right stuff." Rather, it is about what students are *learning* and how we can tell. To review, we have seen that the external world (as represented by employers) has certain expectations of education that turn out to be highly consistent with both learning theories and good practice, as described by cognitive psychologists. In the past decade neurobiologists have provided the biological basis for accepting sound learning theories and practices, while rejecting unsound ones. What does technology have to do with this?

Looking first at the Kolb cycle, we see that computers and calculators can be used as tools to facilitate the CE and AE phases—but not the other two phases, which are right brain and left brain activities. Thus, if the activity allows the student to go directly from CE to AE without engaging the brain—which means the AE will not be *about* anything of substance (except perhaps in the instructor's brain)—it may do more harm than good. Well-designed learning activities involve the entire cycle, at least most of the time.

Now let's consider how technology can support the Seven Principles of good practice described in Section 4.

Student–faculty interaction: Electronic communication via web pages or e-mail enables exchanges with students that never happen in the classroom or office, where students are often intimidated.

Cooperation among students: Here, a lot depends on the atmosphere created by the teacher, but students are quite willing to share with peers what appears on their calculator or computer screen, and this serves as an icebreaker to get them engaged in more substantive conversation. They are somewhat less willing to share their calculations on paper—for which they alone are responsible. If something on the screen does not look "right," there is psychological security in blaming the intervening tool, which is often attributed a personality of its own.

Active learning: This is where technology comes into its own—as already noted for the CE and AE phases of the learning cycle. Students can be much more actively engaged in experiences and experiments with substantial mathematics than was ever possible with pencil and paper.

Prompt feedback: Garbage in, garbage out—immediately. No human teacher can possibly respond as fast or as nonjudgmentally.

Time on task: Students are willing to stay focused on substantial mathematical tasks much longer with technology than without. But there are risks here. If the task is not carefully designed and supervised, students are likely to waste a lot of time addressing questions the teacher never thought of. And, even with well-designed tasks, there is a risk of students subverting the learning process by making their own shortcut from CE to AE, that is, by refusing to think about what they are doing. Even worse, flashy but poorly designed tasks can seduce students into spending lots of time avoiding thought.

High expectations: This is entirely up to the teacher. Technology helps only in the sense that the teacher can expect everyone to carry out tasks that many would be unable to do on their own, e.g., to solve $f(x) = 0$ for some given function.

Diverse talents and ways of learning: Technology is a "great equalizer" in the sense just noted. It's not unusual for a student who lacks skill at symbol manipulation nevertheless to have insight into, say, an emerging pattern of results on the screen. Furthermore, if students are not always told which buttons to push, they will often come up with techniques that are quite different from the instructor's, but just as correct or just as likely to reinforce the correct concept.

7. TECHNOLOGY AND CURRICULUM

For the most part, developers of new curricula have found most of the traditional content still to be relevant, but not necessarily in the same order or with the same emphases or with the same allotment of time. Consider the following example of how technology permits rethinking content and pedagogy to bring them in line with sound theory and good practice.

The raison d'être of calculus is differential equations. Never mind that most calculus students never get there—all the really interesting problems involve ODEs. Traditionally, to understand ODEs one needed lots of technique. And, to get to that point, one needed (in reverse order) power series, integration techniques, the Fundamental Theorem, definite and indefinite integrals, derivatives, limits, functions, trigonometry, and algebra. The traditional curriculum often had a whole semester or more intervening between power series and differential equations, so one really had to be persistent to get to the good stuff.

Now we can pose the problem embodied in a differential equation on Day 1 of a calculus course: The time-rate-of-change of some important quantity has a certain form—what can we say about the time-evolution of the quantity? Furthermore, we can draw a picture of the problem: a slope field. The meaning of solution is then clear: We seek a function whose graph "fits" the slope field. Even the essential content of the existence-uniqueness theorem is intuitively clear—the details can wait for that course in ODEs. By that time, the survivors will have a clear idea of what that course is going to be about and why the details matter.

To be more specific, suppose our initial question is "What can we say about growth of the human population, past, present, and future?" Students recognize that this is an important question, and they start to get engaged with *ideas*. (This is an example of Harel's Necessity Principle.[23]) Students can make conjectures about growth rates, such as proportionality to the population, and explore where they lead. We can trace solutions using the same technique by which we drew the slope field: That's Euler's Method. When we observe that human population is changing more or less "continuously"—1- or 10-year intervals may not be good enough—we are led naturally to the derivative concept. Technology enables experiments that lead to discovery of what's "natural" about the natural exponential function.

There are many models students might pose for population growth, but we don't have to keep guessing. We have about 1000 years of more or less reliable data that we can try to fit to a model. Using a semilog graph, we find that the historic data are *not* exponential. Rather, the growth rate is proportional to the *square* of the population, so the data fit a hyperbola with a vertical asymptote. Students are initially shocked to find that the asymptote occurs within their lifetime (about 2030). Then they really have to think about what all this means. (See Smith and Moore,[24] Chapter 7 Lab Reading.)

The details of the preceding paragraph involve substantial mathematics— numerical, symbolic, and graphical. Note also the echoes of the Kolb cycle: concrete experience with data plots, reflective observation about what the plots mean, abstraction in the symbolic models and their solutions, and active testing of the symbolic solutions against the reality of the data. Then the cycle starts again with the vertical asymptote: What does it mean? How can we fit it into an abstract scheme? How can we test whether our scheme fits with reality?

8. RENEWAL IN CALCULUS COURSES AND BEYOND

It would be foolish to pretend that reformed calculus courses were designed to implement the messages of cognitive psychology or neurobiology. Very few of the developers a decade ago had any knowledge of these subjects. Rather, we had some instinctive ideas about what to try. Some of those ideas were reinforced by our experiences, and they became the basis of our courses. Some ideas didn't work and were quickly forgotten. This is selection at work—but, in order for it to work, we had to challenge our prior knowledge.

Virtually to a person, reformers became committed constructivists, even though few knew that word (in the cognitive sense). In varying degrees, we discovered empirically all seven principles of good practice. The best materials are the ones that encourage students, singly or in groups, to complete the learning cycle—often. The best programs incorporate in some measure all seven of the skill groups identified by employers. And we have learned appropriate ways to use technology to serve learning objectives.*

Departments and individuals who have successfully renewed their calculus instruction have learned—as the National Science Foundation expected—that active learning strategies are not particularly associated with calculus. The same pedagogical principles are equally applicable in courses before, after, and parallel to calculus. Indeed, there are many departments in which, for local reasons, some course other than calculus became the test bed for change.

For many institutions, the next major challenge is to coordinate renewal efforts in mathematics with those of other disciplines—or perhaps to stimulate such efforts. For example, students who have taken reformed calculus courses have often later found quite different (i.e., traditional) expectations in subsequent courses that use calculus. In Bookman and Friedman[26] an electrical engineering major is quoted,

> When I was in Project CALC, it was extremely positive because I wanted to learn the stuff and I had reason to learn it. After Project CALC, none of the other classes that involved any kind of math came easily to me because I was so used to thinking in terms of application.

One wonders what engineering is about if it is not about application.[11] But this experience is neither atypical nor unexpected. It is clear that we still have a lot of work to do with our colleagues in other departments—even in engineering schools that have been highly supportive of reform efforts in calculus.

The problems of science and engineering education were described in Section 3 in the words of an essay by the Pew Higher Education Roundtable.[12] (Recall that *science* here includes mathematics.) The same essay contains constructive suggestions for coping with institutional barriers to change:

1.   Build a research base documenting what works best in terms of making undergraduate science education more accessible and connected.

---

* The reader interested in theories underlying renewed calculus courses, planning and implementing such a course, and assessing student learning should start with Roberts.[6] For additional help on using cooperative learning groups, see Hagelgans *et al.*[25]

2. Restate the mission of science departments to stress the importance of educating a larger population of citizen-graduates to develop a real grasp of science—including those groups that have traditionally been underrepresented in science.
3. Develop strategies and means for establishing operating partnerships that link two- and four-year institutions.
4. Consider ways to foster an effective culture of teaching within departments.
5. Make the quality of science instruction in K–12 schools an explicit priority of undergraduate science education.
6. Develop learning communities that extend beyond the boundaries of individual departments and campuses.
7. Tell science's story out of school.

This is only an outline, of course. Specific actions are suggested in the essay, even for such touchy matters as changing departmental culture.

In spite of successful examples such as Alverno and Evergreen State, the work still to be done on most campuses is daunting. In the words of Pogo, the philosopher of the comic strips, "We are faced with insurmountable opportunities." Can we design collegiate programs in mathematics to meet the needs of our various constituencies—students, their parents, employers, client disciplines, as well as our own profession? Can we do so in a way that is consistent with already identified best practice and with emerging research?

Diamond[13] proposes affirmative answers to these questions, but he notes the importance of integrating individual departments' curricular redesign efforts with those of the entire institution to provide consistent messages and to enable students to practice the core competencies expressed in institutional goals. He writes,

> While all of this is difficult to accomplish, deep curricular change is possible. It does not require adding an exhaustive list of new courses to the curriculum. Many core competencies can be built into required courses. Writing and speaking assignments, activities involving small groups and problem solving, and the development and use of computer skills are instructional techniques that can be introduced or expanded in almost every course. The sequences of instruction must be carefully orchestrated, however, and pedagogy must change.

REFERENCES

1. F. Cajori, *The Teaching and History of Mathematics in the United States*, 1890. Cited by W. Mueller in "The History of Calculus Reform," to appear.
2. A. W. Chickering and Z. F. Gamson (eds.), *Applying the Seven Principles of Good Practice in Undergraduate Education*, in *New Directions for Teaching and Learning* No. 47 (Jossey–Bass, San

Francisco, 1991).

3. A. H. Schoenfeld, "When Good Teaching Leads to Bad Results: The Disasters of 'Well-Taught' Mathematics Courses," *Educational Psychologist* 23(1988):145–166.

4. A. Solow (ed.), *Preparing for a New Calculus*, MAA Notes No. 36 (Mathematical Association of America, Washington, DC, 1994).

5. A. C. Tucker and J. R. C. Leitzel (eds.), *Assessing Calculus Reform Efforts* (Mathematical Association of America, Washington, DC, 1995).

6. A. W. Roberts (ed.), *Calculus: The Dynamics of Change*, MAA Notes No. 39 (Mathematical Association of America, Washington, DC, 1996).

7. A. P. Carnevale, L. J. Gainer, and A. S. Meltzer, *Workplace Basics: The Skills Employers Want* (The American Society for Training and Development and the U.S. Department of Labor, Washington, DC, 1988).

8. *Alverno Magazine*, May 1992. Also see "Alverno's Eight Abilities" (Milwaukee, WI, March 17, 1999), http://www.alverno.edu/glance/g_glance/g_eightabilities.html.

9. Evergreen State College, "We believe" (Olympia, WA, October 5, 1998), http://192.211.16.12/home.ssi.

10. R. E. Boyatzis, S. S. Cowen, and D. A. Kolb, *Innovation in Professional Education: Steps on a Journey from Teaching to Learning* (Jossey–Bass, San Francisco, 1995).

11. Accreditation Board for Engineering and Technology, *ABET Engineering Criteria 2000* (Baltimore, February 22, 1999), http://www.abet.org/eac/EAC_99-00_Criteria.htm#EC2000.

12. Pew Higher Education Roundtable, "A Teachable Moment," *Policy Perspectives* 8(1)(June 1998):1–10 (Institute for Research on Higher Education, Philadelphia).

13. R. M. Diamond, "Broad Curriculum Reform is Needed if Students are to Master Core Skills," *The Chronicle of Higher Education* (August 1, 1997):B7.

14. R. M. Diamond, *Designing and Assessing Courses and Curricula* (Jossey–Bass, San Francisco, 1997).

15. D. A. Kolb, I. M. Rubin, and J. M. McIntyre (eds.), *Organizational Psychology: Readings on Human Behavior in Organizations*, 4th ed. (Prentice–Hall, Englewood Cliffs, NJ, 1984).

16. J. E. Zull, "The Brain, The Body, Learning, and Teaching," *National Teaching & Learning Forum* 7(3)(1998):1–5.

17. M. S. Gazzaniga, *Nature's Mind: The Biological Roots of Thinking, Emotions, Sexuality, Language, and Intelligence* (Basic Books, New York, 1992).

18. J. LeDoux, *The Emotional Brain: The Mysterious Underpinnings of Emotional Life* (Simon & Schuster, New York, 1996).

19. C. Hannaford, *Smart Moves: Why Learning is Not All in Your Head* (Great Ocean Publishers, Arlington, VA, 1995).

20. A. R. Damasio, *Descartes' Error: Emotion, Reason, and the Human Brain*, (Putnam's Sons, New York, 1994).

21. G. M. Edelman, *Bright Air, Brilliant Fire: On the Matter of the Mind* (Basic Books, New York, 1992).

22. R. Sylwester, *A Celebration of Neurons: An Educator's Guide to the Human Brain* (Association for Supervision and Curriculum Development, Alexandria, VA, 1995).

23. G. Harel, "Two Dual Assertions: The First on Learning and the Second on Teaching (or Vice Versa)," *American Mathematical Monthly* 105(1998):497–507.

24. D. A. Smith and L. C. Moore, *Calculus: Modeling and Application* (Houghton Mifflin, Boston, 1996).

25. N. L. Hagelgans, B. E. Reynolds, K. E. Schwingendorf, D. Vidakovic, E. Dubinsky, M. Shahin, and G. J. Wimbish, Jr., *A Practical Guide to Cooperative Learning in Collegiate Mathematics*, MAA Notes No. 37 (Mathematical Association of America, Washington, DC, 1995).

26. J. Bookman and C. P. Friedman, "Student Attitudes and Calculus Reform," *School Science and*

CHAPTER 4

Calculus Renewal and the World of Work

Paul Davis

By one sort of reckoning, 99.9% of calculus students are not deeply devoted to mathematics: Only 1 of each 1000 students who enter the hallowed halls of calculus will ever be ordained with a Ph.D. in mathematics. Accepting a lower level of holy orders as representing true devotion improves the ratios only slightly: Just 0.5% of calculus students will earn either a master's or a doctorate in mathematics. Even if mere acolyte status—declaring an undergraduate major in mathematics—represents an adequate profession of faith, fully 93% of calculus students still remain in outer darkness.

Proof: In the fall of 1997, about 455,000 students enrolled in first-year calculus courses in the United States. At the same time, approximately 59,000 juniors and seniors had declared themselves to be mathematics majors. During the preceding academic year, 1174 doctorates in mathematics had been granted, 522 to U.S. citizens.[1] Supposing these numbers to be invariant over time, that only U.S. citizen doctoral recipients studied calculus in the United States, and that junior and senior undergraduate majors were evenly divided between the two class years leaves the approximate ratios 522:29,500:455,000, or 1.15:64.8:1,000, of Ph.D. to mathematics major to calculus student. Approximately five master's degrees are awarded for every doctorate in mathematics.[2]

Such sarcastic estimates are easy to criticize, for their arrogance if nothing else. But the fundamental point is unavoidable. Most calculus students are not in it for the mathematics. Of course, the same can be said of students of introductory physics or chemistry or history or literature. How many students of English I aspire to become the next Danielle Steel?*

What becomes of the 99% of calculus students who will not do what we do, who will not teach college or university mathematics? Some indeed will become teachers, perhaps at a precollege level, perhaps in a discipline other than mathematics. But most will pursue what my nonacademic friends pointedly call *real work*, a career in business, industry, or government. For purely selfish reasons (the alumni fund and a solvent social security trust, among others), we should all hope that they are equipped to be productive and, if it suits their aspirations, well paid.

What does work outside the academy require of our calculus students? To what degree do those demands coincide with the goals of calculus renewal? What did the calculus reform movement ask of itself, and what might continued calculus renewal demand in the future? How do those aspirations mesh with the requirements of employment? The surprising discovery is the extraordinary agreement between the goals of calculus renewal and the expectations of many employers, a congruence that offers mathematics a stronger role in education and a more certain place in society at large.

1. THE MANDATE OF CALCULUS REFORM

The calculus reformers have yet to convene a Council of Trent to define their essential dogmas.† However, the Tulane conference of 1986 did outline goals for calculus renewal, both through explicit statements and by contrast with the then current state.

Apropos the world of work, among other worlds, Peter Lax suggested to the participants in the Tulane meeting

> Without [applications] a calculus course is in danger of resembling a guided tour through a carpentry shop, with instructions on how to use each tool [including some antiquated ones], but giving no sense of how to use them to build a thing of beauty and utility; or a music class where most of the time is spent in practicing scales and finger exercises, with little chance to listen to or play a composition much above the level of chopsticks; or a language class where grammar and syntax are taught systematically, but where there is little conversation, composition or reading of literature.[3]

Of course, "applications" are areas of work in which mathematics is believed to be useful, and music, language, literature, and the rest are fields in which some

---

* On the other hand, no student of English would question the importance of reading, much less confess to illiteracy. But being "math negative"—being innumerate—carries only slight social stigma, even for students of calculus.

† In the context of calculus reform, the orientation of this metaphor is reversed. The Roman Catholic Church convened the Council of Trent in response to the Protestant Reformation, not vice versa.

people work and others browse merely for enjoyment and stimulation. An optimist might read into Lax's wide-ranging metaphors the suggestion that calculus instruction has a multitude of roles to play.

Lynn Steen posed to the conferees more than 20 questions, some of which may yet beg answering.[4] Among them are

- Is calculus an appropriate filter for the professions?
- Do students really learn the major ideas of calculus?
- Is calculus a good course to train the mind?
- Does calculus contribute to cultural literacy?

His questions also suggest that introductory calculus carries a large burden, both for the world of work and for the world of the mind (a distinction that college and university faculty have the luxury to avoid).

The Methods Workshop at the Tulane conference did offer a set of recommendations that sound suspiciously like goals.[5] Among them are

- To help students understand a select set of fundamental topics in calculus, and ... to apply what they have learned with flexibility and resourcefulness
- To expose students to a broad range of problems and problem situations, and a broad range of approaches and techniques for dealing with them
- To help students develop an appreciation of what mathematics is, and how it is used
- To help students develop precision in both written and oral presentation
- To help students develop their analytical skills and the ability to reason in extended chains of argument

These recommendations retain the broad charge at which Lax and Steen hinted, but certain attributes are now made explicit, including *flexibility, resourcefulness, broad range of approaches and techniques, written and oral presentation, analytical skills, and ability to reason.* Many of these characteristics are required in the careers pursued by students of calculus.

2. THE DEMANDS OF THE WORKPLACE

The academy values knowledge as well as the observation and contemplation by which it is acquired. Business and industry value analysis as well as the synthesis and action which result. Henry Pollak has commented on the consequence for mathematics of that distinction[6]:

> There is a rich mathematical experience intertwined in many jobs, but, in many cases, that experience is not part of the mathematics that we teach or that our students learn.

Brockway McMillan has identified the reason why so little of the experience of nonacademic work involves the mathematics we teach: "Mathematical solutions can be found only for mathematical problems."[7]

Of course, outside of cliques of mathematicians and their intellectual allies, common discourse seldom uses the formal language of mathematics. Hence, few of the problems encountered in the world at large are posed as mathematical problems.

To learn something of the work environment for the 99.5% of calculus students who do *not* earn an advanced degree in mathematics, one might begin with what is known of the nonacademic workplace for those who *do* hold a Ph.D. or a master's in mathematics. Presumably, workers whose formal training in mathematics ended at a lower level will face *non*mathematical demands at least as intense as these highly trained workers, such nonmathematical demands compensating for the lower level of disciplinary training.

In fact, common themes emerge in studies of several related disciplines—computer science,[8,9] economics,[10] and mathematics[2]—once issues centered on specific disciplinary expertise are set aside. The experience of computer science and its interactions with the broader field of information technology are instructive. A quick tour of employment and training issues in computer science offers mathematicians the opportunity to view relatively dispassionately a situation similar to their own.

Computer science programs train computer programmers (though the majority of them actually come from other disciplines[11]), and computer programmers are widely thought to be in short supply.[12] However, computer programmers, narrowly defined, are in some sense a commodity item, and many segments of the information technology community caution against training with such a limited goal, for it does not lay a foundation for future learning.[9] Indeed, careful reviews of the latest reliable data suggest that the strongest industrial demand in the information technology field is for workers with higher-level skills, those who are able to synthesize and create, to lead and manage complex projects, not just exercise a narrow set of technical skills.[11]

Such conclusions are no surprise to members of the computing community who had absorbed an earlier National Research Council report[8] on the state of the computing work force. In answer to the question "What do employers want?," it had listed such responses as

> [T]echnical competency is definitely not enough.

> [Y]ou need ... people who are malleable, open to change ...

> [We need] good communication skills.

That report goes on to outline the importance to industry of working in multi-disciplinary teams, and it carefully distinguishes industrial development from academic research. For example, in reference to the former, one manager said, "We don't do research... We often break new ground in applying old technology to new problems" (p. 51).

Academic members of that report panel[8] cautioned that such "a focus on generic skills may come at a price, since individuals with advanced training ... offer a more sophisticated approach to solving technical problems" (p. 44). Naturally, mathematicians would offer the same caution.

Mathematics departments do not produce commodity workers like computer programmers. A few semesters of the product rule and implicit differentiation will not train individuals with marketable skills like those of entry-level C or Java programmers. But absent an analog of such "production-line" programmers and with an allowance for technical differences, the demands placed on mathematicians in business, industry, and government are otherwise parallel to those encountered by computer scientists. In addition to technical expertise, mathematicians are also expected to exhibit higher-level skills of creativity and innovation in ways that do not necessarily mimic academic research.

The nature of employers' expectations (and much else about the workplace) for individuals with advanced degrees in mathematics has been described in the report *Mathematics in Industry*[2] from the Society for Industrial and Applied Mathematics (SIAM). Carefully structured telephone interviews and follow-up site visits, ultimately involving about 500 individuals, led to the conclusion (p. 1) that

> [s]ome of the most important traits in nonacademic mathematicians include
>
> - skill in formulating, modeling, and solving problems from diverse and changing areas;
> - interest in, knowledge of, and flexibility across applications;
> - knowledge of and experience with computation;
> - communication skills, spoken and written;
> - adeptness at working with colleagues ('teamwork').
>
> The qualities that distinguish these mathematicians from other scientists and engineers are seen by their managers as falling into two broad categories:
>
> - highly developed skills in abstraction, analysis of underlying structures, and logical thinking;
> - expertise with the best tools for formulating and solving problems.

The SIAM report makes clear that the environment for mathematics in industry is not that of the university. There are few mathematics departments. Doctoral mathematicians find themselves in workgroups in which, on average, only 25% of the members are mathematicians; master's mathematicians constitute only 16% of their workgroups. In the workplace, unlike the academy, mathematicians do not work solely with their coreligionists.

A significant majority of the managers who participated in the SIAM study were not mathematicians. Comments from those managers and from participants in the site visits give a sense of the fundamental values of the industrial workplace.[2]

> Managers evaluate their people by what they contribute to the company.

> The most effective people are those who can interact, understand, translate. A key is being able to explain something outside your discipline.

> Lack of interpersonal and team skills is the primary cause of failure in industry.

> A metaphor for success that we heard from more than one manager was the letter T, which meant that a successful mathematician must have depth in an area of specialization but at the same time develop a broad understanding of technical and business issues in the company.

> Once in an industrial position, many mathematicians find that meeting the demands of their organizations' mission takes precedence over their disciplinary identities.

> The hardest task for a mathematician is developing the real problem requirements. The user doesn't usually know what the solution will look like in the end.

> Mathematicians here must extract the mathematics from the projects that need it.

> Problems never come in formulated as mathematics problems. A mathematician's biggest contribution to a team is often an ability to state the right question.

> Most problems must be "solved" in hours or days; this often means finding an adequate solution rather than a perfect one.

Given such management attitudes and the typical graduate program in mathematics, it comes as no surprise that graduate mathematicians felt their educational preparation was "less than good" in such area as the following (SIAM,[2] p. 20):

- Working well with colleagues
- Communicating at different levels
- Having broad scientific knowledge
- Effectively using computer software and systems
- Dealing with a wide variety of problems

These observations apply to holders of advanced degrees. Of course, a bachelor's graduate will possess less well developed skills of abstraction, will have a more limited toolkit for formulating and solving problems, and will have followed a broader curriculum. Those differences notwithstanding, corroboration of the need for such abilities and attributes as computing, problem solving, flexibility, communication skill, and teamwork can be found both in other formal studies and in anecdotes about the work experiences of individuals with less specialized mathematical training.[13,14]

For example, holders of bachelor's degrees in mathematics who work in nonacademic positions spend one-third of their time with computer applications (Table 5 of Ref. 2), while more than half the doctoral graduates in the SIAM study rated computing as "essential" (SIAM,[2] p. 14). Their occupations include a good many areas that are not part of mathematics curricula, such as sales and marketing or management (Table 3 of Ref. 2), substantial evidence of the need for cross-disciplinary collaboration.

The Mathematical Association of America's *Career Profiles*[13] collected accounts of career experiences from individuals with various levels of mathematical training, including some who were not even undergraduate mathematics majors. Some of their comments put a personal face on the range of requirements of the world of work, particularly the importance of personal relationships, the requirement of productivity, the ubiquity of computing, and the challenge of working across a range of specialties and problems:

> Being able to translate business situations and problems into a mathematical model is an invaluable skill to possess in the corporate world. However, the full mathematical contribution can be achieved only by using "people" skills to obtain a clear understanding of the problem from the customer and then to interpret the "numbers" solution. (M. Reynolds in *Career Profiles*[13])

> My mathematical training allowed me to adapt very quickly and to be productive almost immediately. (C. Beeman in *Career Profiles*[13])

> Without my math foundation, I would not be competent to design roads, drainage systems, flood relief projects or water and sewer systems. (D. Farmer in *Career Profiles*[13])

> I started with the company . . . as a computer programmer. [Later] I designed computer software for business systems such as: accounts payable, time management, steam collection from turbine generators, sales systems, inventory management and many others. . . . [Now as] Corporate Technical Developer . . . I often [must] . . . evaluate new technologies. (J. Cooper in *Career Profiles*[13])

Given these perspectives on the role of mathematics in nonacademic jobs, are we to confine our teaching only to that mathematics that is fully explicit in the workplace or only to that mathematics that is isomorphic to what we do? If so, what becomes of the 90% or more of calculus students who will *not* be working in a job whose title includes *mathematics*?

Although the context may differ and the problems may not be overtly mathematical in nature, we may be certain that skill in formulating, analyzing, and solving problems will be required by the future employers of our calculus students; that flexibility in meeting new business opportunities will be essential; that communication ability will be fundamental; and that collaboration with co-workers will be taken for granted.

Does the calculus curriculum have a role in fostering these attributes?

3. OLD TOOLS FOR A NEW CHALLENGE: RECOMMENDATIONS

Of course, the narrowest interpretation of the responsibilities of teaching calculus would reject as irrelevant such challenges from the world of work. For most mathematicians, and for a small minority of our calculus students, "[mathematics] *is* sufficient."[15] But for many students, the connection to the workplace is valuable motivation for mathematics, and it provides important training in life skills. A new challenge for calculus renewal is making that connection while still making good mathematics.

In some respects this challenge is old news. Recall some key words (italicized here) from the objectives set by the Tulane conference:

- ... *understand* ... fundamental topics ... apply ... with *flexibility* ...
- ... *broad range* of problems and problem situations ... *broad range* of approaches and techniques ...
- ... an appreciation of what mathematics is, and how it is *used*.
- ... precision in ... *written* and *oral* ...
- ... develop ... *analytical* skills ... ability to *reason* ...

The challenges of the workplace are indeed not new, although the point of reference may be. Perhaps unconsciously, reform calculus has already been addressing many of these needs through now-familiar techniques. For example, the Rule of Three—examine each idea from three perspectives, analytical, graphical, and numerical—is an exercise in flexibility (as is the "Rule of Three Plus One,"[16] which adds the physical perspective). Problem solving fosters analysis and is itself a core skill. Pedagogical methods like active learning and group work have blossomed in parallel with the renewal movement, and they require teamwork while teaching communication. Finally, computing is as ubiquitous in most reform calculus courses as it is in the workplace.

Perhaps then the challenge posed by the 90% or more of calculus students who will not become mathematicians is really quite modest. Meeting that challenge requires recognizing connections to the workplace and explicitly accepting some share of responsibility for nurturing them. A consolidation of current ideas and approaches for renewing calculus, coupled with a commitment to the needs of the world of work, would allow mathematics instruction to shoulder a new responsibility for which it is particularly well suited, the training of creative thinkers whose contributions to economic welfare will arise directly from their modern education in mathematics. Consciously structuring our teaching to acknowledge the importance of mathematical patterns of thought, even if those patterns fall outside the formal boundaries of mathematics itself or fail to use its vocabulary, would better motivate the majority of our students in the short run while educating a new generation in the crucial importance of mathematics.

Three recommendations summarize the challenges for those of who teach calculus:

1. Learn more about the world of work outside the academy and the role of mathematics in it.
2. Choose sound pedagogical methods that teach mathematics while also building such workplace skills as breadth and flexibility, teamwork, and communication.
3. Accept the role of mathematics in helping students grow as learners within the university and later as productive and responsible members of society.

3.1. Learn More about the World of Work

The first recommendation is not difficult. Our own graduates who are working in business, industry, and government are a ready source of information, and most are happy to return to campus to speak with us and our students. Likewise, campus recruiters are ready to tell us of their business needs, particularly if such conversations might lead them to new, well-qualified employees. Placing students in internships and cooperative employment or incorporating into the curriculum problems and projects with off-campus sponsors is a more significant change, but the compensation is an even deeper, firsthand understanding of the needs of the workplace, coupled with new motivation for mastering mathematical ideas.

3.2. Choose Sound Pedagogical Methods that Teach Workplace Skills

This recommendation might be dismissed by some as an invitation to "vocational education." If *vocation** is understood in its dictionary meaning, "a regular occupation, especially one for which a person is particularly suited or qualified,"[17] then there should be little shame among us for making our students "particularly suited" for careers that use mathematics. If the phrase is misused as a slur to connote mindless exercise of memorized skills without understanding or insight, then the objection is misplaced, for no teacher wants such drudgery to pass as education and no employer wants to hire college graduates who approach their jobs so artlessly.

The benefit of teaching mathematics with an eye on the workplace is the development of generations of students who can credit their mathematical training with some portion of their professional success, even if they are working

---

* *Vocation* is derived from the Middle English *vocacioun*, a divine call to a religious life, a curious metaphorical coincidence.

in areas far removed from formal mathematics. Conversely, mathematics will remain at the margin in large parts of society as long as mathematical thinking is undervalued in the workplace.

3.3. Accept the Role of Mathematics in Helping Students Grow as Learners

The third recommendation offers mathematics faculty a new way to think about their mathematics teaching: couple the implicit workplace demand for lifelong learning with conscious development of basic academic skills, the fundamental tools of continuous learning. Since the first calculus course is often one of a new student's first encounters with higher education, calculus could strengthen its place in the curriculum by intentionally nurturing some of the core skills required by both university education and employers.

For example, a calculus course could consciously improve communication skills (reading, writing, speaking, and listening), foster working in groups with unfamiliar individuals, and practice synthesizing new information by assigning group projects that required mastering some material outside of mathematics (e.g., to permit an application of optimization ideas) and culminated in an oral or written report. In collaboration with experts from other parts of the campus community, such an assignment could incorporate focused instruction in some of these basic skills or it could be coordinated with such instruction elsewhere in the curriculum. With such approaches, basic calculus could become a foundation for lifelong learning and productive, worthwhile work rather than remain for so many students a random obstacle in the college curriculum.

This recommendation asks more than simply teaching calculus, but perhaps not much more when it is valued in terms of return on investment, a return paid to our students and to our discipline. Building academic and workplace skills within the calculus course offers mathematics the opportunity to pioneer a renewed and broadly interdisciplinary understanding of the role of service courses in university education. Such courses can consciously address the development of skills that aid learning during the college years and promote professional growth after graduation. The discipline that is bold enough to stake out such a role for itself will be more highly valued within the university because it is better serving both its students and its intellectual foundations.

For some of us, mathematics is sufficient. But for many of our calculus students, perhaps not 99.9% but certainly the vast majority, more is necessary. The next decade of calculus renewal should preserve teaching that builds mathematical understanding and traditional disciplinary values. But it should augment that teaching with the development of skills that will weave mathematics into our national culture through the workplace.

REFERENCES

1. P. W. Davis, J. W. Maxwell, and K. M. Remick, "1997 AMS–IMS–MAA Annual Survey (Second report)," *Notices of the American Mathematical Society* 45(9)(1998):1158–1171.
2. Society for Industrial and Applied Mathematics, *The SIAM Report on Mathematics in Industry* (SIAM, Philadelphia, 1998) Table 16, p. 31. Available at http://www.siam.org/mii/miihome.htm.
3. P. Lax, "On the Teaching of Calculus," in R. G. Douglas (ed.), *Toward a Lean and Lively Calculus*, MAA Notes #6 (Mathematical Association of America, Washington, DC, 1986) 69–72.
4. L. A. Steen, "Twenty Questions for Calculus Reformers," in R. G. Douglas (ed.), *Toward a Lean and Lively Calculus*, MAA Notes #6 (Mathematical Association of America, Washington, DC, 1986) 157–166.
5. R. B. Davis *et al.*, "Report of the Methods Workshop", in R. G. Douglas (ed.), *Toward a Lean and Lively Calculus*, MAA Notes #6 (Mathematical Association of America, Washington, DC, 1986), xv–xxi.
6. H. Pollak, *The Algebra Initiative Colloquium*, C. B. Lacampagne, W. Blair, and J. Kaput (eds.), US Department of Education, Office of Educational Research and Improvement, National Institute on Student Achievement, Curriculum, and Assessment (Government Printing Office, Washington, DC, 1995).
7. B. McMillan, "Applied Mathematics in Engineering," in U. Dudley (ed.), *Readings for Calculus*, v. 5, MAA Notes Volume 31 (Mathematical Association of America, Washington, DC, 1993) 150–155.
8. National Research Council, *Computing Professionals: Changing Needs for the 1990s*, Steering Committee on Human Resources in Computer Science and Technology, Computer Science and Telecommunications Board (National Academy Press, Washington, DC, 1993), p. 42 ff.
9. P. Freeman and W. Aspray, *The Supply of Information Technology Workers in the United States* (Computing Research Association, Washington, DC, 1999).
10. W. L. Hansen, "Report of the Commission on Graduate Education in Economics," *Journal of Economic Literature* 24(1991):1035–1053; "The Education and Training of Economics Doctorates," *Journal of Economic Literature* 24(1991):1054–1087.
11. B. S. Barnow, J. Trutko, and R. Lerman, *Skill Mismatches and Worker Shortages: The Problem and Appropriate Responses*, 25 February 1998 draft final report to the U.S. Department of Labor (The Urban Institute, Washington, DC), p. 45.
12. *Help Wanted: The IT Workforce Gap at the Dawn of a New Century* (Information Technology Association of America, Washington, DC, 1998). [See the careful review of this report in Ref. 11.]
13. Mathematical Association of America, *Career Profiles*, http://www.maa.org/careers/index.html.
14. American Mathematical Society–Mathematical Association of America–Society for Industrial and Applied Mathematics Project for Nonacademic Employment, *Mathematical Sciences Career Information*, http://www.ams.org/careers/. [Provides a variety of information about nonacademic employment, including a few profiles of mathematicians working in industry.]
15. U. Dudley, "Is Mathematics Necessary?" *The College Mathematics Journal*, 28(5)1997:364.
16. P. W. Davis, "Asking Good Questions about Differential Equations," *The College Mathematics Journal*, 25(5) (1994):394–400.
17. *The American Heritage Dictionary of the English Language, Third Edition* (Houghton Mifflin, New York, 1992).

CHAPTER 5

Technology and Calculus

Wade Ellis, Jr.

1. INTRODUCTION

The use of computers–both desktop and handheld—has gradually changed the learning and teaching of calculus over the past decade, at least in textbooks. Calculus textbooks now routinely have graphing calculator exercises and many have computer-based exercises and projects. Few textbooks come without ancillary packages including computer test banks, computer laboratory manuals, multimedia CD-ROM-based learning aids, Web sites, and mathematical word processing software. How students learn and how professors teach calculus, however, has not changed much. Nor has much research been done on the effect these technology-related changes can have on student performance in mathematical contexts. This chapter will focus on what students and teachers might do differently in mastering calculus in a technology-rich environment and also what researchers could investigate to improve the learning and teaching environment where technology is readily available.

The chapter will discuss:

- Factors that may affect how the calculus curriculum will change
- Presenting mathematics
- Tools for mathematical computations
- The current impact of technology on course content
- The future impact of technology on course content
- Learning at a distance

- Issues surrounding faculty development
- Conclusions and recommendations

2. FACTORS THAT MAY AFFECT HOW THE CALCULUS CURRICULUM WILL CHANGE

Many mathematicians believe that computers have changed the world, but not the world of mathematics. Thus, there is no need for change in teaching calculus. Some would argue, however, that computer hardware and software have changed the nature of mathematics in much the same way that the introduction of Vieta's symbolic representations of variables and constants changed mathematics from a primarily geometric field based on figures (concrete) to a study based on symbols (abstract). This change soon caused geometry to be deemphasized—and, much later, resulted in the demise of the secondary school geometry course in the mid-1960s. Specifically, this course became little more than many careful symbolic definitions such as an angle as the union of two rays with a common endpoint (that few students understood and none cared about). Computers have changed the practice, if not the nature, of mathematics. What mathematicians do has changed, how they prove theorems has changed, how they make conjectures has changed because of the billionfold increase in numerical and symbolic computation and in numerical computations leading to graphical representations. If the practice has changed, perhaps the courses taught and how they are taught should change as well.

Second, the number of topics in calculus courses (all developed before the advent of the mathematical software) has all too frequently prevented instructors from entertaining any change. The advances in cognitive theory, however, should change our approach to learning mathematics and to learning how to apply mathematics in other disciplines (if the two can be separated). Scientists are taking advantage of this new knowledge to create inquiry-based courses in chemistry, biology, and geology. In part, the widespread availability of Internet access to scientific sites containing large collections of interesting and useful data has made this possible. Some in the mathematics community have experimented with such inquiry-based approaches. Perhaps the advances in cognitive theory can affect calculus courses as well.

Finally, the underlying contentiousness of undergraduate mathematics faculty in the United States has impeded innovation. The ad hominem attacks and counterattacks involving those who would move toward an inquiry-based approach to mathematics in the classroom, necessarily discounting centuries of blackboard-and-chalk-based experience in presenting mathematics courses to students, are particularly intense because the calculus curriculum is seen as embodying the essence of mathematics. Thus, a discussion of how to rework ideas and methods to improve student understanding very quickly becomes a pitched battle over small differences in fundamental principles. Some hold that the content and methods of calculus as the gods (Euler, Cauchy, Gauss, Weierstrass, Riemann) presented it are sacred; how dare anyone to change them. They told their students what to think, we will tell ours as well. Students cannot discover anything until they have fully mastered the knowledge and skills we will reveal to them. Such approaches to teaching were good enough to solve Fermat's last theorem and put a human being on the moon. Maybe they are not as outdated as some reformers might think. The mathematics community might remember that the gods changed the then existing mathematical landscape. Perhaps the current landscape needs to change as well.

3. PRESENTING MATHEMATICS

3.1. In the Classroom

There are a number of computer software packages that allow instructors to present information rapidly and effectively in class. These packages have a variety of excellent features, but perhaps the most important of them is the ability to use a computer to present crisp text and graphics in an interactive mode where the instructor can switch back and forth among screens with accurately drawn diagrams whose order has been carefully thought out in advance. The instructor is relieved of chalkboard management and can concentrate on insightful explanations and interactions with the students based on the displayed screens. Although much of the advantage of such presentations could be attained with slides or transparencies, the ease of creating colorful and inviting visual presentations and the ability to move quickly to a particular diagram without fumbling through a stack of transparencies is substantial. In addition, it is much easier to rework lectures that are stored and presented in electronic form. Also, such presentations can be easily printed out in a reduced format that allows students to take supplemental notes without redrawing graphs and diagrams.

Some presentation packages such as Mathematica, Maple, and Scientific WorkPlace also allow the instructor to perform computations accurately, smoothly, and swiftly at the time of the presentation. This allows for an added level of interaction with the students who can suggest different approaches to a particular problem. Most instructors are not immediately comfortable with these capabilities, but with time they learn how to create interesting and instructionally effective courses. Timing each diagram or slide, knowing how much information to place on each slide, knowing how to most effectively direct student attention to parts of the slide, and fostering student questions in such an environment is initially a tall order even for the most energetic instructors.

3.2. Research opportunities

Although there is anecdotal evidence that such presentations are very effective in the hands of masterful teachers, research needs to be done on how these methods affect student performance and student attitudes toward mathematics. Do carefully prepared chalkboard lectures have the same effect as computer-based presentations? Is the interactive nature of such technology-based presentations actually realized or does the technology place yet another barrier to student understanding and successful performance?

3.3. On the Desktop

The March 1999 issue of the *SIAM Review* (SIAM, Philadelphia, Volume 41, Number 1, pp. 133–163) has a new section entitled *Education*, which presents supplemental material for applied mathematics and scientific computation intended to be read by students. Authors are encouraged to have references to Web sites that provide multimedia supplements.

From the introduction to this new section:

> ... it will provide modules that teachers and students can use directly in studying applied mathematics and scientific computation. All modules will be available on-line as well, sometimes including links to additional, supplementary materials.

This new approach to presenting journal articles points the way for a multimedia reworking of all the important topics in calculus-related fields (which contain much of the mathematics that is applied) based on presentation packages like those discussed above. If this were to occur, apart from supplying steady employment for legions of mathematicians who are underemployed at the moment, it would have a profound impact on what mathematics is taught and how. If the notion that we cannot teach our students all the mathematics they need to know to be successful during their professional careers gains favor, then perhaps what we need to do is give them the opportunity and guidance to be successful in mastering mathematics presented in this new way suggested by the *SIAM Review*.

3.4. Research Opportunities

There are numerous possibilities for research that come to mind in such a new environment. Do the existing examples of such modules work? If so, what are the characteristics of students who are successful in learning from such modules? Can the characteristics needed to learn such modules be developed by specific learning activities? If so, what are those activities? What are the properties of successful modules that students are able to learn from? Can these properties be taught to authors or are they particular to specific topics?

Are there topics that cannot be taught in this way? If so, what are the properties of topics that can be presented in this way? What curricula might be effective in encouraging students to develop the skills, knowledge, and attitudes that will allow them to succeed in learning from such modules? Finally, are there social conditions that foster learning in such environments?

4. TOOLS FOR MATHEMATICAL COMPUTATIONS

Digital computers have performed mathematical computations for the last half-century. Software now performs numerical, symbolical, and graphical mathematical computations on inexpensive ($100) and very expensive ($1,000,000) devices throughout the mathematical community. The list of software packages and hardware/software calculators includes Maple, Mathematica, Derive, MathCAD, MatLab, Axiom, and the TI-89 and the CASIO CFX-9970G. Such computer software is capable of performing all the standard algorithms of calculus, in most cases instantaneously. The mathematics teaching community has attempted to address this technology in many ways. First, as with most new technologies, we attempted to do the things that we were currently doing using the new software. We used the software in demonstrations in class to show what the computer could do. Then we progressed to allowing students to use the software to do some of the more difficult computations with the help of the software in laboratory activities or projects. Finally, we have begun to allow students to use the computer to actually do the mathematics as it occurs in the course. Each of these activities has been successful at some level and unsuccessful at others.

4.1. Traditional Calculus Textbooks

When computer time was expensive, demonstrations in class or just displays of computer printout were effective because they announced to the student that computers existed to perform mathematical tasks and they did so with as little expense as possible. The laboratory activities were successful in that they allowed instructors to see a variety of mathematical activities as valid in providing students with interesting learning environments where students could grow intellectually. Finally, the use of the computer as an integral part of the computational apparatus that students could use caused the mathematics community to rethink what they were teaching in the calculus course. At the moment, the calculus course is in disarray. The leading textbooks in the marketplace admit to allowing any of a variety of courses to be taught from them. An elementary real analysis course, an inquiry-based learner-centered project-oriented course, a computer-algebra-system-based course, an applications-based course, a hand-computations-based course—all these courses are a possibility from these text-books. If the student is to read such books, what part of them should she pay attention to? What part should she ignore?

4.2. Reform Calculus Textbooks

On the other hand, the reform (some would say revised) curriculum calculus texts that rely heavily on new approaches to classroom activity are too narrowly focused to allow for instructor preference. To use them, the instructor must understand the authors' point of view, agree with it, and not deviate from the presentation by omitting, adding, or rearranging material. One might characterize such books as technology-friendly because they allow for the use of technology as an investigative tool (as do the most recent traditional curriculum textbooks), but they do not look to changing the curriculum based on the new technology available to the students. Neither type of textbook truly embraces the technology.

5. THE IMPACT OF TECHNOLOGY ON COURSE CONTENT

In such a loaded environment, questions about the nature of mathematics, about pedagogy, about the uses and abuses of technology, about the underlying basic concepts of calculus, about the goals of the calculus course, and about the type of student now coming to calculus, all intersect and cause confusion. In what follows, the discussion will be restricted to the effect that the technology has had and can have on the content of the course itself.

5.1. Techniques of Integration

The use of technology has put at risk the "techniques of integration" topic. Almost every textbook does more with integral tables, but little with the use of software in finding antiderivatives. Not many textbooks emphasize the idea that few functions have closed form integrals, although much is done with hand-drawn abstract functions with multiple discontinuities. Exact answers are empha-sized, but the approximation of definite integrals whose value is theoretically known to exist but cannot be found using an antiderivative is not. Nor are bounds on such approximations often emphasized.

5.2. Graphing Functions

Most textbooks discuss the use of graphing software and calculators in graphing functions with the first and second derivative tests. But still much time is spent on exact sketches (an oxymoron) of pathological functions. Every

textbook warns against software errors, but few encourage students to use software as a trained but fallible assistant in mathematical work to verify results in as many ways as possible—from simple mental estimation to hand computations and sketches and software assisted computations.

5.3. Multivariate Calculus

An effective use of technology in calculus occurs in many textbooks on multivariate calculus that contain computer-drawn graphs of functions of two variables and contain problems and projects that require the use of Maple or Mathematica for computations or Cyclone or MatLab for three-dimensional graphing. These activities have greatly increased student understanding of difficult mathematical material and allowed students to create multiple visual images of concepts that were previously only vaguely understood with the help of a few static textbook diagrams and graphs. Unfortuately, only a small proportion of calculus students take this course.

5.4. Research Opportunities

These changes in the way topics are presented have at one level embraced the technology and used it effectively. These uses give rise to a variety of research questions. Do students understand a definite integral as a number better through the use of such software? Do students have a better understanding of anti-derivatives through the use of such software? Do students understand the fundamental theorem of calculus more thoroughly by using such software? How much more do students learn about functions and their graphs through the availability of instantly drawn graphical images of functions at many different magnifications? Do the multiple dynamic images of functions of several variables enhance student understanding of partial derivatives, of multiple integrals, and of the various coordinate systems used in three-dimensional graphing?

6. THE FUTURE IMPACT OF TECHNOLOGY ON COURSE CONTENT

But there is another level of software use in the calculus curriculum that the mathematics community is just beginning to explore. Just as geometry became less important in "school and university" mathematics as the power of algebraic symbol manipulation became clearer and its use more widespread, will some of the topics of calculus become less emphasized as the power of numerical, graphical, and symbolic capabilities of computer software are better understood and more widely used? It is difficult to believe that the current calculus curriculum is only a half-century old and was not a gift from the gods. In fact, the curriculum is based on the blending of differential and integral calculus around the fundamental theorem of calculus that George Thomas of MIT (*Calculus and Analytic Geometry*, Addison–Wesley, Reading, MA, 1951) developed just after the Second World War and replaced the clear delineation between differential calculus and integral calculus of the Granville, Smith, and Longley approach (*Elements of the Differential and Integral Calculus*, Ginn, Boston, 1904) that reigned for the first half of the twentieth century. Perhaps it should be noted that each of these textbooks is and was in print for over 50 years and that the problem sets in the early editions were quite short. These textbooks were successful in the United States as they provided the mathematical foundation for scientists and engineers who developed nuclear fission and fusion, jet propulsion, digital computing, and computer science.

The notion of the antiderivative and the ability to find antiderivatives is required to solve differential equations in closed form, but the Riemann sum approach to the definite integral is not a requirement. Since finding antiderivatives is basically reversing (where possible) the process of differentiation, it may be possible to go directly from an understanding of the derivative as a rate of change to writing and solving differential equations without the long development of the definite integral as a Riemann sum. In this context, we might wish to view the fundamental theorem of calculus as an existence theorem that guarantees the solution to the differential equation $y' = f(x)$, where $f(x)$ is a continuous function.

6.1. A New First-Year Collegiate Mathematics Curriculum

Putting these notions together with the idea that most functions do not have closed-form antiderivatives and, thus, most differential equations do not have closed-form solutions, one might attempt to create a different kind of introductory first collegiate course in mathematics based on the numerical, graphical, and symbolic computational capabilities of computer software that focuses on:

- The understanding of basic ideas (limit, derivative, the fundamental theorem of calculus)
- The importance of theory to guarantee that solutions to differential equations exist, and
- A strong reliance on real-world problems to motivate the study of approximation and the idea of limit

Such a course would begin with difference equations as an introduction to the study of change and to show the limitations of finite approaches to change. The course could then develop the derivative as a method of measuring change and the differential equation as an investigative tool in furthering the study of change. Although symbolic solutions to differential equations would be emphasized, a clear indication of the limitation of symbolic methods would also be presented. Symbolically intractable differential equations would be investigated using the numerical and graphical methods that have become accessible to students through inexpensive and powerful mathematical software. The accuracy of numerical methods and their limitations would be discussed as well.

A course based on this outline would provide students in disciplines that have a one-year calculus prerequisite with the capability of using difference and differential equations as powerful investigative tools. Students would also develop a working knowledge of specific computer software packages and the kinds of difficulties that arise in using such software indiscriminately, without a theoretical and practical understanding of its mathematical foundations and the context in which it is used. Such a course might better serve the needs of our client disciplines, be more interesting to their students, and spark a renewed interest in all students in the usefulness of mathematics, its beauty and power, and its attractiveness as an area of study.

6.2. Research Opportunities

Many research questions surround this approach to the first year of collegiate mathematics that is now called calculus. What student performance is important in a first-year collegiate course in mathematics? What student performance measures do we wish to develop? What are the effects on student performance of wholesale use of a computer algebra system like Maple, Mathematica, or Derive? How much does this use improve student symbol sophistication (and how do we define such sophistication)? How much does such use improve student inventiveness in attacking problems? How much does this use improve the ability of the student to solve routine problems? Nonroutine problems? What basic mental and paper-and-pencil manipulative skills are needed before students can appropriately use computer software (topic by topic and overall)? How does the importance of definitions, theorems, and proofs change with the introduction of software that embodies the theory students are learning? How do student attitudes toward mathematics change with the use of such tools? How do student attitudes toward theoretical considerations and justifications of statements, conjectures, and theorems change? Does symbolic computational software raise yet another barrier to student understanding of mathematics and mathematical situations? In what ways are students' mathematical capabilities enhanced with the use of computational software and in what ways are they left undeveloped by such use?

7. ISSUES SURROUNDING FACULTY DEVELOPMENT

A variety of uses for computers in calculus-level courses have been presented. These uses are expanding through the mathematics teaching community, though slowly. The spread of the use of technological tools into the college curriculum is much more rapid in other scientific disciplines because such use is driven by the research community. Whereas recombinant DNA experiments are required in the first collegiate course in biology, the use of software in doing mathematics is a debatable endeavor and in many cases is forbidden in the first course in collegiate mathematics.

There are clearly some good reasons for the reluctance of the mathematics community to embrace this technology. The axiomatic–deductive approach to mathematical truth is an enduring aspect of mathematics that is appropriately cherished by all mathematicians. Students have difficulty mastering the set of ideas that underlie such an approach (the same is true for the scientific method). This approach requires a level of mental rigor and toughness that students seem reluctant to attain. Computer software may prevent students from acquiring a mathematical mindset that demands proof of each and every step, of each and every statement. Software can hide the steps and the justifications. Many students believe that whatever the computer "says" must be correct.

The mathematics teaching community must develop a greater understanding of what the technology can do, what its theoretical underpinnings are, what pitfalls and difficulties it represents, and how best to take advantage of its enormous power in mathematics research and mathematics education. To do this, we must first develop a knowledge of the software.

7.1. Obstacles to Learning about Mathematical Software

An obstacle many faculty face in learning about software is the folk wisdom that software (and its uses) really is not mathematics—though by-hand computation is. Thus, many teachers in colleges and universities have never used a computer algebra system, solved a differential equation (or any equation) with such a system, or seen the kinds of ridiculous—but correct—results that machines can generate. For example, the implementation of Ferrari's algorithm for the general solution to the quartic equation in Maple, Mathematica, or Derive can go on for several pages. The solutions are correct, but it is impossible to tell if they are real or complex, rational or irrational, positive or negative.

There are other obstacles. Many teachers of college-level mathematics believe that mathematics is (largely) computation—in contradiction to the idea that software computation is not mathematics. Thus, a machine that appears to do the computations is usurping the role of the mathematician. That the increased power of computation allows more—and more varied—examples so that more mathematics can be done by more people more of the time cannot occur to them. Also, many instructors lack practical experience in the use of mathematics. Many teachers of calculus believe that using the textbook techniques of integration to find antiderivatives is an effective way to evaluate a definite integral of a function,

that it works most of the time. Though they possess the theoretical knowledge that almost all continuous functions do not have closed-form antiderivatives, they lack the practical experience to apply this knowledge in working with real-world problems. Thus, they believe in the appropriateness of teaching difficult and mind-numbing special case symbol manipulations to find antiderivatives that can be instantaneously computed by software.

7.2. A Critical Issue

Faculty development is, therefore, critical to the implementation of appropriate uses of technology in the teaching of calculus. Such faculty development should be approached on two levels. First, instructors must be encouraged to understand what software can do and what the limitations of software are. Then they must be encouraged to use this knowledge in developing new approaches in old courses and new courses that use this knowledge to the benefit of the students and the mathematics community. This can be accomplished in two steps:

1. Obtain (well-paid) faculty summer internships, available at many national and private laboratories (Argonne National Laboratories, IBM), scientific research organizations (SRI International, Xerox Palo Alto Research Center), government agencies (National Security Agency, Department of Energy), scientific research firms (Genentech Inc., Merck & Co.), computer hardware and software firms (Intel Corp., Cisco Systems Inc., LucasFilm Ltd., Microsoft Corporation), and financial institutions (Credit Suisse First Boston, The Prudential Insurance Company of America). The internships should last for several summers and might require that interns go in pairs from their home institutions.
2. Secure reassigned time during the academic year, in which the interns conduct workshops at their home institutions for their own departments and for other interested faculty about what they learn during their internships and how it can be applied in courses that they are teaching.

As these interns spread their knowledge of the real world through their own departments and client discipline departments, students who will become tomorrow's teachers will see mathematics at work in a technology-rich environment where real-world problems are investigated. This is likely to increase the number of mathematics majors, as well as the number of able students who are interested in mathematics. In this way, mathematicians can begin to change the culture of academic mathematics toward an interest in society and the world. In addition, this process would create entry-level mathematics instructors who know about technology, value its use, are able to use it well, and are willing to develop new courses that use technology effectively.

Some will say that this is a long process. Remembering that it took nearly 50 years to create the current environment, one would be surprised if it took only a few years to change it.

7.3. Research Opportunities

From a research point of view, faculty development efforts present many interesting questions. Does the Thomas Kuhn theory of scientific innovation carry over to teaching innovation as well? What are the characteristics of faculty members who are willing to participate in internship programs? What percentage of the faculty will allow itself to be retrained in on-campus workshops directed by a colleague? What are the characteristics of faculty members who will accept retraining? What are the most effective means (types of workshops) to retrain them? Is fear the only stimulus to change in the faculty? How much is a faculty member's allegiance or loyalty to the discipline of mathematics likely to hinder the development of new teaching materials based on the usefulness of mathematical knowledge to the society? Do professors see pedagogical considerations as important in introducing computer software into courses?

8. LEARNING AT A DISTANCE

No discussion of technology would be complete without some attention to distance learning. The movement to distance learning has been limited because no college, Internet company, or publisher has learned how to make it profitable. As soon as the correct environment exists for profit making, there will be a wholesale move toward distance learning. To some extent, this has been prefigured by (usually nonprofit) scientific journals which, viewed as educational tools, are based on the notion of learning at a distance. The new *Eduation* section of *SIAM Review* mentioned earlier attempts to enhance this type of learning. The College Board Advanced Placement Examinations are also a form of distance learning as each is a syllabus with an examination that is graded away from the student (who has learned the course at school or at home). In some ways a textbook is distance learning. A student working alone learns material that has been developed by someone else at a distance from the student (both physically and intellectually).

However, the typical definition of most distance learning in mathematics involves some form of multimedia activity (videotape, slide presentation, simulation, animation, interactive computer software, or a combination of these) that can be accessed by individuals working alone at home or in a school environment. Currently, the fact that most distance learning students work alone clearly limits the potential effectiveness of such programs. And, very few individuals possess the motivation, time management skills, and persistence to work through such materials alone. Most currently or soon to be available commercial products assume some access to other students and an intructor. For example, Academic Systems of Mountain View, California, suggests that students view their multimedia courses in a laboratory setting with other students with whom they can collaborate and with a knowledgeable person who can answer questions if learning or equipment problems cannot be resolved. Textbook publishers have developed multimedia packages, some of which have the ability to obtain personal assistance via e-mail using Web sites designed for their packaged courses. Some of these packages allow students to work together over the Internet using the computer screen as a whiteboard. However, these systems suffer from the difficulty of conveying mathematical symbols accurately and easily across the Internet. This is a severe limitation not experienced in multimedia courses for many other disciplines.

8.1. Online Assessment

One major advantage of distance learning with the appropriate programming is that each student's knowledge and skill level can be assessed regularly and, based on such assessments, appropriate activities for that individual student can be provided. The software available from the ALEKS (Assessment and LEarning in Knowledge Spaces) Corporation in Irvine, California, may have precalculus modules available by the spring of 2000 that assess a student's knowledge state from over 40,000 such states selected for a particular knowledge space like precalculus. It then provides the student with information about the precalculus skills and ideas the student knows well, ones the student knows (but not well), and what topics (based on what the student knows) the student is able to learn next. The ALEKS software then presents the student with the option of choosing one of these "knowable" topics to work on at the computer. The ALEKS system can continually assess the student and provide the appropriate educational materials over the Web until the student has mastered the knowledge space.*

Such software allows a college or university to more accurately assess student deficiencies prior to a calculus course and provides remediation before or while the course is in progress. This can alleviate some of the need for remedial courses, provide instructors with students who are appropriately prepared for courses, and reduce student frustration in learning material that is based on material that they do not remember or never learned. Such software, if appropriately used, can increase student retention rates and student satisfaction with their education, especially at institutions with diverse populations (both geographically and demographically).

---

* The Web site for the ALEKS Corporation is currently www.aleks.uci.edu.

The expense and complexity of using such packages at colleges and universities are daunting, but some institutions are making a start. Perhaps the best way to use such packages is in a hybrid fashion. Algorithmic skills can be introduced in a classroom setting and then practiced using the software multimedia package. Mathematical thinking, problem solving, and an appreciation of mathematical approaches to problems and situations can continue to be the responsibility of the classroom teacher. This approach maximizes the time for the instructor to guide, encourage, and instruct the student while ensuring plenty of practice time with fundamental skills. The future of higher education on college campuses may well be determined by the ability of mathematicians to create such hybrid courses that attract students to the classroom because of the need for human interaction. At the same time, instructors need to be aware of the varying needs and preparation of students and how such hybrid assessment-based courses can satisfy students' needs—at least in part—because of their computer-based assessment capabilities.

8.2. A Calculus Curriculum for Every Discipline

One ongoing assumption of the mathematics teaching community is that there should be a monolithic engineering and science calculus curriculum where the "real calculus" is taught. Industrial organizations, especially automobile firms and clothing firms, have determined that each customer should have exactly the product that she wants and computer technology has made it possible to give it to her. The same may be true for mathematics. Can technology be used to tailor calculus courses for specific groups of science and engineering students? Can there be a special course for demography students, for management science students, for chemistry, for physics, and for biochemistry students? Can this be done using distance learning, since no one of us can learn all of these different disciplines? Can the technology be tailored to enhance such calculus courses? In management science, spreadsheets could be used. In engineering, it could be MathCAD or MatLab. In chemistry, Mathematica. Perhaps there would be a core course with Internet laboratories in the individual disciplines. There are many possibilities. Perhaps mathematicians should be more creative and student-oriented in their approach to the ever more specialized, discipline-based, mathematical requirements of many (if not most) of the students in undergraduate mathematics courses.

8.3. Research Opportunities

Since the College Board, Educational Testing Service, and others are developing multimedia calculus materials for those secondary school students who do not have access to an Advanced Placement mathematics course, it is expected that calculus will soon be offered through distance learning courses, some of which will be very high quality. Such courses give rise to a myriad of research questions. What are the characteristics of students who succeed in such distance learning courses? What are the characteristics of those parts of the course that are most successful based on the voluminous test data that are available, say, from the Advanced Placement multiple-choice and free response tests? How do students who take these courses perform in subsequent courses? What attitudes do students who take distance learning calculus courses have toward mathematics? How well do students learn algorithms in such distance learning courses? How well-developed are their problem-solving skills? How well do they perform on problems that are nonroutine? What type of distance learning activities are preferred by students: animation, slide presentations, videotaped lectures, interactive software? What types of distance learning activities are most effective— animation, slide presentations, videotaped lectures, interactive software—as evidenced by student performance? How much do students use Internet-based student interactions?

9. CONCLUSIONS AND RECOMMENDATIONS

Technology is changing the way calculus is taught and learned, as well as the topics presented and the interactions in and out of the classroom. Each of these changes requires training, evaluation, and a constant rethinking of the goals in teaching students calculus. We no longer live in an environment where there are many textbooks that are the same and which are based on a textbook template that will last 50 years. Much is changing; much will change. Each instructor will be faced with learning anew how to meet students' needs and demands, while remaining true to mathematics and to the core human values of the academy in a technology-rich environment filled with charlatans and wonderful and challenging new ideas.

The mathematics community will need to consider, discuss, and debate the changing nature and practice of mathematics, what the fundamental values of the community are or should be, and how the essence of mathematics can be maintained while the uses of mathematics become more varied, important, and technology-driven. These discussions must result in changes to the educational environment in colleges and universities that will support a complex society that is based on the use of logical thought and technology in every aspect of daily life.

SUGGESTED READING

1. W. C. Bauldry and W. Ellis, Jr., *Calculus: Mathematics and Modeling* (Addison–Wesley Longman, Reading, MA, 1999).

2. P. Blanchard, R. L. Devaney, and G. R. Hall, *Differential Equations* (Brooks/Cole, Pacific Grove, CA, 1998).
3. W. E. Boyce and J. Ecker, "The Computer-Oriented Calculus Course at Rensselaer Polytechnic Institute," *The College Mathematics Journal* 26(1995):45–50.
4. "Computer as Electronic Blackboard: Remodeling Organic Chemistry," *Educom Review* (Spring 1991):35 (with S. L. Cassanova).
5. J.-P. Doignon and J.-C. Falmagne, *Knowledge Spaces* (Springer-Verlag, Berlin, 1998). Also available from J.-C. Falmagne, Department of Cognitive Sciences, University of California, Irviine, CA 92607.
6. H. Gardner, *The Minds New Science: A History of the Cognitive Revolution* (Basic Books, New York, 1985).
7. M. B. Hayden, *Newton: An Interactive Environment for Exploring Mathematics* (University of Rhode Island, 1996). Dissertation Abstracts International 57(1997):3434A.
8. M. K. Heid, "Resequencing Skills and Concepts in Applied Calculus Using the Computer as a Tool," *Journal for Research in Mathematics Education* 19(1988):3–25.
9. P. T. Judson, "Calculus I with Computer Algebra, "*Journal of Computers in Mathematics and Science Teaching* 9(3)(1990):87–93.
10. A. W. Roberts, *Calculus: The Dynamics of Change* (Mathematical Association of America, Washington, DC, 1996).

CHAPTER 6

Renewing the Precursor Courses

New Challenges, Opportunities, and Connections

Sheldon P. Gordon

1. INTRODUCTION

The earth's crust is composed of a series of large plates floating on the underlying molten magma. These plates are constantly shifting and consequently they bump into one another and often one will ride up on top of another near their edges. The interfaces between the plates form fault lines in the earth and the resulting pressures that build up along the interfaces eventually release to form earthquakes.

The mathematics curriculum can be viewed in much the same way as being composed of a series of such plates. Historically, the two largest—and perhaps the most important—are the secondary and college mathematics curricula. Over the last several decades, the college-level plate has split into two, one of which includes the traditional first two years from calculus through differential equations and linear algebra and the other includes the developmental courses designed to replicate the secondary-level curriculum for either students who avoided mathematics in high school or students who need to learn the material again. However, there are associated plates in other disciplines, such as physics, engineering, biology, and economics, as well as one for upper-division and graduate level mathematics. All of these plates affect the plate for the first two years of college mathematics.

For most of our careers, the primary mathematical plates—developmental, the first two years, and the upper division—have been quite stable. The underlying magma on which they ride has been solidified, so there has been relatively no movement. In particular, the interface between the first two plates has been quite smooth. The high school curriculum was well established and mathematicians were very cognizant of its content, having experienced it firsthand. Consequently, mathematicians could count on its invariance in the development of curricula at the college level. When we designed courses that are precursors to calculus—developmental algebra, college algebra, and precalculus—they were essentially clones of the high school equivalents, though sometimes with the proviso to do it "faster and louder."

A major part of the comfort level with this curriculum is that mathematicians know what it does well—by replicating the experiences they went through, this curriculum produces clones of themselves. Unfortunately, only a very small percentage of the students in these courses today have the capability, or the desire, to fit that mold. And worse still, the value placed on people with this kind of training is diminishing. At the same time, these courses do not provide the mathematical needs of the other students; there is a good reason why, nationally, enrollment in mathematics drops by about 50% in each successive course from ninth-grade algebra on up.

Now, the entire structure of mathematics education is in the process of change. The postsecondary curriculum is undergoing a series of dramatic changes. The calculus reform movement has made some significant and far-reaching changes. As a result, consequent changes are needed in the subsequent courses, such as multivariate calculus and differential equations, as well as in the precursor courses that prepare students for calculus. There are growing pressures to increase students' exposure to statistical ideas and probabilistic reasoning early in the curriculum; likewise, there are pressures to introduce matrices sooner. What then of the interface between the secondary curriculum and the college curriculum or that between the precursor curricula that exist at the college level and the mainstream curriculum?

2. CHANGES IN THE UNDERGRADUATE CURRICULUM

Think about the overall structure of the courses that prepare students for calculus in the traditional mathematics curriculum. In Algebra I we teach most of the fundamental rules and methods of manipulative algebra. While *we* teach it, *they* don't learn it. Therefore, at least 80% of Algebra II is devoted to repeating the same topics. So, why does the follow-up course in college algebra, or precalculus for that matter, still reteach the same rules and methods of algebra? Because, although we still teach it, most of the students still haven't learned it. Of course, there are always a few exceptions—those who are exceptionally good at the manipulations. The mathematicians teaching these courses are among those exceptions.

There was a lot of truth in the old adage "you take calculus to learn algebra." In traditional calculus, students were routinely expected to *use* algebra—often for the first time—to solve substantial problems or for lengthy derivations. It was no longer just a meaningless game of drill for no apparent purpose; if students couldn't do the algebra, then they couldn't do calculus. Unfortunately, the emphasis in too many calculus courses shifted to endless drill and practice on long lists of derivatives and integrals. Thus, in the process of finally learning algebra, most students came away with the perception that calculus primarily consists of facts such as the derivative of x^3 is $3x^2$. Relatively few students ever understood what the derivative of a function really meant, even though they could differentiate quickly. So there was the corollary to the above adage, "you take differential equations to learn calculus."

The calculus reform movement insists that students should learn and understand the fundamental concepts of calculus while taking calculus. The reform projects place far greater emphasis on conceptual understanding of the ideas of calculus, achieve a better balance among graphical, numerical, symbolic, and verbal representations, and incorporate more realistic applications, often from the point of view of mathematical modeling with differential equations.

The changes in calculus, while originally envisioned as being primarily content changes, quickly included pedagogical changes as well, with frequent emphasis on more active learning environments (such as collaborative and cooperative learning), use of individual and group projects, and emphasis on writing and communication. In order to accomplish this, however, there has been a concomitant reduction in the traditional emphasis on symbolic manipulation. Most of the instructors teaching these courses believe that the courses are the better for these changes since they focus on calculus and its value; they believe that the students are the better for it because they are learning calculus and how to apply it; and they feel that they personally are the better for it, since they are finally teaching mathematics and not simply more algebra. (Of course, there are certainly some who believe that a de-emphasis on algebraic manipulations is a dreadful loss.)

The changes in both the content and pedagogy in calculus have set the stage for comparable changes in differential equations and other postcalculus courses. Simultaneously, the new calculus courses are also having an impact at the secondary level as the Advanced Placement (AP) calculus course has changed to reflect the same ideas and goals. Finally, comparable changes are working their way into those college courses that were the traditional calculus preparatory track (see Solow[1]). A variety of reform projects have resulted in alternative courses at the precalculus, college algebra, and developmental algebra levels that likewise place less emphasis on many of the routine algebraic skills, particularly those associated with factoring polynomials and operations with rational expressions. Instead, these new courses focus more on conceptual understanding of the fundamental mathematical ideas—variable, function, behavior of functions—as well as on realistic applications of the mathematics. They often feature "new" mathematical content, such as the use of real-world data (once thought to be the domain of statistics) and the notion of fitting a function to the data, or aspects of probability, or recursion and iteration, or substantial applications of matrix algebra that go well beyond solving systems of linear equations.

In many ways, this reflects the new paradigm for the mathematics that is actually used in practice. In several presentations, including one at the U.S. Military Academy (West Point) conference on the Core Curriculum (see Dossey[2]), Henry Pollak has used an illustration such as that in Fig. 1 to dramatize this shift away from the mathematics used by practitioners (engineers, scientists, applied mathematicians, and others who used the mathematics) in 1960, say, when most of today's senior mathematics faculty were in school. At that time, virtually every problem considered by the practitioners was continuous and was approached from the point of view of seeking a closed-form, deterministic solution. Relatively few problems were discrete in nature; some were stochastic in the sense of having a random component. The mathematics curriculum of the time typically mirrored this paradigm closely.

Today, a very different paradigm exists in terms of the mathematics used, as shown in Fig. 1—virtually every real-world problem now has a major discrete component, if it is not inherently discrete; even continuous models must be discretized to permit computational solutions. Most realistic problems today have a random component—there is always some degree of uncertainty. Yet, the mathematics curriculum today does not reflect these new emphases. For the most part, the content is the same as in 1960, although there is slightly more emphasis

1960 **Mathematics**		2000 **Mathematics**	
Discrete Deterministic	**Continuous Deterministic**	Discrete Deterministic	Continuous Deterministic
Discrete Stochastic	Continuous Stochastic	**Discrete Stochastic**	Continuous Stochastic

Figure 1. The changing face of the mathematics used in practice.

on statistical reasoning and some discrete topics now exist in the curriculum. One of the major challenges mathematicians face is integrating more of these ideas and methods into the curriculum at all levels (see Dossey[2]). In the courses that are precursors to calculus, the larger challenge is to do this while preparing students for calculus.

At the same time, these precursor courses are increasingly unsuccessful for a variety of reasons that will be discussed later. An analysis of data in a report by the Conference Board on the Mathematical Sciences[3] indicates that perhaps 10% of the students who take college algebra or precalculus courses in college ever go on to calculus. Many students take these courses to fulfill a quantitative literacy requirement. Many students planning to continue do so poorly in these courses that they "leak out" of the mathematics pipeline. Perhaps it makes sense to rethink the intent of such mathematics courses and to make them valuable in their own right, rather than to think of them as courses designed exclusively to prepare students for calculus.

3. CHANGES IN THE SECONDARY CURRICULUM

While most college-level mathematics faculty have naturally focused on the changes that are being felt at the college level, very significant forces have been transforming the secondary curriculum as well. If anything, this transformation has been happening for considerably longer than those at the college level. Most college faculty have been aware of them, but only peripherally. We decry the fact that incoming freshmen appear to have poorer manipulative skills and less of the information that has traditionally been considered important for success in college level mathematics. Based on what can be inferred from dealing with these students and based on our own high school experiences, most college mathematics faculty typically conclude that either the students are academically worse or that the high schools are completely at fault. In turn, the colleges have reacted by introducing and expanding their developmental and other precursor offerings to bring the students up to speed. At many institutions, these precursor programs are the majority of the mathematics offered, with more sections, more students, and more institutional resources expended than in any other area.

Perhaps, though, the problem really results from a shift in the secondary curriculum plate, implying that the smooth interface we have always expected is no longer there. All mathematics faculty have heard about the *Curriculum Standards* from the National Council of Teachers of Mathematics (NCTM),[4] but few have paid great attention to them and far fewer have ever read them. Yet, the *Standards* is having an ever-increasing impact on *what* is taught in the secondary schools and *how* it is taught.

The *Standards* call for a very different approach to teaching mathematics, emphasizing a greater depth of understanding of mathematical concepts and reasoning. They also emphasize geometrical and numerical ideas as a balance to purely symbolic ideas, as well as the use of substantial applications of the mathematics, often in the context of group projects. The *Standards* require increased communication on the part of students in the form of written and oral reports. They often involve collaborative learning activities. They also presume that technology assumes an appropriate and ongoing role in the teaching and learning of mathematics and, in reality, the use of graphing calculators is quickly becoming universal since they were mandated for the AP calculus exam beginning in 1995 and are allowed on almost all other standardized tests. The *Standards* calls for the early introduction of many different mathematical ideas into the curriculum, particularly statistical reasoning and data analysis, matrix algebra and its applications, recursion and iteration, and probability. Overall, implementation of the *Standards* implies higher expectations for teachers and students. And, in actuality, these goals and expectations match closely the goals and expectations of the reform efforts at the college level.

However, something has to be removed to make room for all these new emphases. In the process, therefore, the *Standards* call for a diminished emphasis on traditional formal algebraic manipulation. For example, students spend considerably less time factoring polynomials. Instead, students are expected to understand the notion of the roots of an equation, which they apply to factor simple expressions. They learn how to determine the real roots of more complicated equations graphically and numerically and then use these roots as needed. Presumably, these students will have far more understanding of the concept of a root; however, the cost is that they may have somewhat less facility at recognizing factors at a glance.

Is this a fair trade-off? *In principle*, most faculty at the college level will welcome students with such backgrounds. Most of these changes are completely compatible with the spirit of the calculus reform movement and with the changes that are beginning to be implemented in courses before and after calculus. *In practice*, however, things are somewhat different. All of the mechanisms that currently bridge the relatively small gap between the secondary school and the introductory collegiate plates now have to be rethought.

These comments also assume that such new courses exist widely in the secondary schools. How extensive are they in actuality? From discussions with individuals who have developed materials for such courses, it is estimated that at least one-third of all new textbooks purchased by the secondary schools reflect new curricula. The figure may be much higher, particularly as new editions of traditional texts incorporate more of the reform ideas. In addition, many teachers are implementing some of the ideas and themes in the absence (mostly for budgetary reasons) of formal sets of textbooks.

4. COLLEGE PLACEMENT EXAMS

One of the most critical areas at the college level that needs rethinking is that of the placement exams used. A large and growing proportion of the students coming in from high schools have been exposed to very different mathematical ideas and emphases, yet mathematics departments continue to assess their ability and knowledge on the basis of a curriculum that is rapidly disappearing. It is little wonder that so many students seem to place lower and lower on these exams despite having had two, three, or four years of secondary mathematics. It may not be that they have failed to learn what they were taught, but rather that they were taught different ideas and methods from those assessed on traditional placement exams. Perhaps it is time to look at both the currently existing secondary curricula and at college-level placement procedures to see if there is any connection between the two. It is quite likely that, at most colleges, many entering students will reflect a traditional curriculum; many other students, however, will reflect a different curriculum. Thus, there will likely be a need for multiple versions of placement tests, one for students with traditional backgrounds and one for students with backgrounds that include coursework based on the NCTM *Standards*.

But this raises a series of additional problems. For example, when teaching a reform calculus course, one of the hardest challenges for many instructors is creating test questions that are in the spirit of the course. The same applies when teaching reform courses prior to calculus. In addition, different instructors bring different perspectives to the types of nontraditional questions they ask; there certainly is no universal consensus about what is "standard" in reform courses. What then should go into a placement exam that measures what students have learned in reform courses? It is relatively easy to compose questions that measure how well students have mastered and retained traditional algebraic skills. It is considerably more daunting to attempt to measure the level of their conceptual understanding, for instance. Just deciding what to measure will be a challenge. In addition, it is questionable to what extent standard placement exams measure a student's readiness for reform calculus or any other reform offering, regardless of the flavor of the student's secondary school mathematics preparation. Therefore, it is becoming increasingly imperative for new placement tests to be developed, either in-house or by the professional test providers.

5. THE ROLE OF TECHNOLOGY

Technology has tremendous implications for the teaching and learning of mathematics, particularly in the precursor courses. The problem educators face is how to use the available technology to support the teaching and learning of mathematics, rather than as an end in itself.

The use of graphing calculators or computer graphics certainly allows one to introduce a much more geometric flavor to almost all mathematics courses. Instructors no longer need to dwell on procedures for producing the graphs of functions, but can instead emphasize the behavior of different classes of functions and focus on questions of scale (i.e., domain and range) to get reasonable pictures that highlight the important characteristics of the functions. Rather than developing a large variety of techniques designed to solve carefully crafted equations, instructors can look at the general issue of solving any equation in one variable. If the equation is particularly simple, students should be able to solve it using pencil and paper. But, if an equation is more complicated, or does not possess a closed-form solution, students can "solve" the equation graphically by zooming in on the solution point or they can solve it numerically through the table features included with most calculators. If a Computer Algebra System (CAS) is available, they may be able to solve it algebraically as well.

And, if less time is spent on routine mechanics, it becomes possible to introduce other topics. For instance, it is now possible to consider more realistic applications because we are no longer limited to solving equations having simple closed-form solutions. Alternatively, one can introduce a variety of more "modern" topics—such as fitting functions to real-world data, which is one of the most important applications of mathematics in practice, or recursion and difference equations, which can serve as a preparation for differential equations.

One of the major themes in reform calculus is that of the multiple representation of the function concept, so that students see and understand the interplay between symbolic, graphical, and numerical approaches, as illustrated in Fig. 2. This theme can be even more powerful in the precursor courses, and

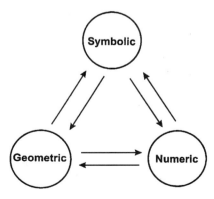

Figure 2. The interplay among symbolic, geometric, and numerical representations.

students should understand how to go from any one of the three representations to the other two perspectives for any function. In the past, mathematics courses were highly focused on going from the symbolic representation of a function to construct its graph manually by first constructing a table of values. In practice, the reverse is by far the more standard problem—given a table of data values or a graphical display, construct a function that models the data. That function can then be used to answer questions that naturally arise in the context as opposed to solving artificially created equations for the sake of practice.

Now that CAS capabilities are available on relatively cheap graphing calculators, there is an even greater challenge for mathematics and a concomitant potential for change affecting the entire mathematics curriculum. Widespread adoption of such a level of technology requires a total rethinking and restructuring of the developmental and college algebra courses to significantly reduce the time spent on pencil-and-paper manipulation skill. In turn, if there is less need to repeat the drill on the same skills from one course to the next, as was previously discussed, then the potential exists to shorten the precursor experience by one or more semesters. However, there are some major drawbacks that could accompany such a move. Students going through such a program might not develop any mathematical maturity, at least not in the senses that mathematicians traditionally recognize. Also, a CAS is useless if one does not use it wisely. There is little doubt that there will be a variety of schools that will experiment with such large-scale programs, with inevitable successes and failures. The mathematics community should pay close attention to the experiences of these experiments, because they may have much to say about the future of the entire mathematics curriculum.

However, even with the availability of such CAS technology, it is evident that students will never be as fast or as accurate as that calculator or computer program—no matter how much instructors drill or motivate their students. Instead, it is essential that faculty adopt a higher goal to produce something far more valuable than an imperfect organic clone of a calculator or computer program.

6. RETHINKING THE PRECURSOR COURSES TO CALCULUS

Mathematics education is currently at a very remarkable stage in history where, because of the confluence of all the pressures and changes, the rare opportunity exists to reassess all undergraduate offerings. Does it make sense to offer a full spectrum of precursor courses, from developmental algebra up through precalculus, that are "clones" of traditional high school courses? That depends on the institution. At schools that have implemented calculus and other reform programs, it makes no sense to take students who have had secondary school courses offered in the same spirit as the reform programs and enroll them in precursor courses that are primarily symbolic manipulation to prepare them for calculus courses that no longer treat symbolic manipulation as the primary emphasis. But what then of the students who have had traditional secondary courses? How can the college curriculum help them transition to the reform courses?

On the other hand, colleges that offer more traditional calculus courses face a very different problem. They may still have to offer standard precursor courses, but possibly from a very different point of view. Specifically, it is not necessarily the case that students who take such courses are "poor" or "weak" mathematically, but just that they have had a nontraditional mathematical background. They shouldn't be viewed as "remedial" students, but rather as students who may need a different approach to the mathematics to bring them up to the desired level.

For that matter, many mathematicians assess students' mathematics ability entirely in terms of their facility with symbolic manipulation: those who can do algebra quickly and correctly are "good" students and those who cannot are "poor" students. But there is much more to mathematical ability. This point became dramatically clear to the present author a year or so ago during a college algebra class while leading the students to discover and enumerate the behavioral characteristics of cubic polynomials. One of the students, a young woman who was a high school dropout, raised the seemingly innocuous question: "Is it true that every cubic is centered at its point of inflection?" Seeking to draw her out, I asked, "What do you mean by that?" With her eyes screwed up as she tried to visualize her image and with her hands moving in opposing directions, she responded, "Well, if you start at the point of inflection and move in both directions, don't you trace out the identical path?" This demonstrates an incredible level of mathematical insight, especially for someone who had an extremely low view of her own mathematical ability based on teachers' perceptions and her own performance in traditional secondary courses. Incidentally, I had discovered this lovely fact myself only a year or so before and have since found that many mathematicians do not seem to be aware of this result.

However, based on the reform approach that many students are now getting in secondary courses, it is little wonder that they find traditional precursors terribly boring. They have been exposed to much richer mathematical experiences—mathematical ideas and substantial (as opposed to artificial) applications and problems. It should come as no surprise that traditional precursor courses are increasingly unsuccessful. It is not that the students are weaker mathematically, but rather that many have come to expect more from their mathematics courses. If faculty want to develop a traditional level of manipulative skill, they will have to rethink how to motivate students. And that rethinking will likely produce precursor courses that reflect many of the same themes as the reform courses at the secondary level.

In addition, most of these students will come to college having used some technology (usually a graphing calculator) in their secondary mathematics courses. Now that the SAT and AP exams allow graphing calculators, it should be expected that the majority of incoming college students will bring a graphing calculator with them. It seems unreasonable to these students that they are not allowed to use their calculators (or other more sophisticated technology) in college courses. Many secondary mathematics teachers respond to such complaints from their ex-students by advising their current students not to apply to colleges that do not permit use of technology.

7. THE IMPORTANCE OF MANIPULATIVE SKILLS

In today's world, far more people regularly use the concepts of calculus than sophisticated manipulative algebra. People must be able to interpret graphs and tables. They must understand the concept of a functional relationship and how to use it intelligently to make predictions. They need to understand relative growth and decay rates (the real use of percentages); they must know about increasing and decreasing rates of growth (concavity). They must be aware of the notion of accumulation. Many must be aware of the notion of parameters and how changes in parameters affect the behavior of a process. Most need to know something about the modeling process.

But very few people need to factor something like $x^8 - y^8$, let alone $\cos^8 t - \sin^8 t$. Very few must be able to reduce

$$\frac{\left(\dfrac{x_2 - 4x + 3}{x^2 - 5x + 6}\right)}{\left(\dfrac{x^2 + x - 2}{x^2 - 4}\right)}$$

to the obligatory 1. In fact, other than while teaching algebra courses, how many mathematicians ever actually perform algebraic operations as complicated as these? It is easy to picture an academic research mathematician whose teaching load consists exclusively of liberal arts courses or introductory statistics along with virtually any upper-division or graduate course who will never need to use any sophisticated algebraic skills. Equally important, much the same can be said about most engineers and scientists. It will be increasingly true in the future as both their training and their practice of engineering and science depend more heavily on technology.

This certainly does not mean that there is no longer a need to teach algebra and that students should not be expected to perform any algebraic operations whatsoever. However, there likely will be a very different balance. Certain algebraic techniques will continue to be taught and, if anything, are likely to be emphasized more than in the past. For instance, operations with properties of exponents and logarithms fall into this category because they are needed to solve many of the problems that are receiving greater emphasis. But many other skills, particularly those associated with solving equations, will receive far less attention. For instance, it is less important to be able to factor a polynomial to find its roots if one can locate any real root to any desired degree of accuracy by graphical or numerical means. Similarly, applying an inverse trigonometric function finds just one possible solution to a trigonometric equation, but all solutions can be found graphically. Also, it is far more time-consuming (as well as limiting in terms of the degree of complexity expected of students) to solve a system of linear equations by hand when it can be solved by converting it to a vector–matrix equation and pushing the appropriate keys on the calculator; this operation simultaneously provides an exposure to the power of matrix algebra and its applications.

A redefinition of the word *algebra* is now needed, one that has far-reaching implications for the courses that teach it. Algebra should be viewed as far more than just a collection of manipulative tools for moving symbols around and for solving carefully constructed equations. This is especially true in today's fast-changing world. The traditional precursor courses were designed primarily to develop algebraic skills that once were essential for success in later courses. The wide availability of technology and changing requirements, especially in the client disciplines, requires a rethinking of this paradigm. For instance, students in upper-division courses in engineering and the sciences do relatively little pencil-and-paper mathematics; instead, they focus on developing mathematical models to describe real-world phenomena. These models typically involve differential or difference equations, matrix methods, or often probabilistic simulations. The students examine the behavior of the solutions, particularly as the parameters underlying the phenomena change. Similarly, students in business, the social sciences, and the biological sciences are expected to recognize trends from sets of data, construct appropriate mathematical models to fit the data, and make corresponding predictions based on the models developed. This is actually remarkably similar to what students in lab courses have been doing for centuries; the difference is that the students in the business and social science courses typically use spreadsheets for the analysis rather than hand-drawn graphs.

Unfortunately, in the minds of many students, there is little connection between what they see in mathematics classes, where the problems look like: *Find the equation of the line through P* (1,2) *and Q* (4,8), and what they encounter in their other courses, where the problems are more like: *Given a set of points, draw the line that best fits the points, find its equation, and answer the following questions about the situation from which the data came.* Faculty in other quantitative disciplines complain that the mathematics taught is too abstract.

Mathematicians tend to interpret this as a complaint about too much rigor: too much definition–theorem–proof structure; too much delta–epsilon, and so forth. In reality, what the other faculty are communicating is that the methods used in mathematics courses are context-free and idealized to the level of sterility. Students do not recognize that they need to apply the same ideas and techniques to problems that vary even slightly from the textbook examples. For example, when the students see situations where the letters used for the variables are other than x and y or the numbers used are not one-digit integers, they often have no idea that there is any connection at all with mathematics.

In general, the overarching emphasis on algebraic manipulation in traditional precursor mathematics classes does not provide the foundation that students need for these disciplines, nor does it adequately prepare them for reform calculus. Instead, a broader preparation is needed, one that better reflects the practice of mathematics. Students need to learn how to

1. Identify the mathematical components of a situation (i.e., model it)
2. Select the right tool (e.g. paper and pencil, graphing calculator, CAS package, spreadsheet) to solve the problem
3. Interpret the solution in terms of the original situation and, if necessary, change the assumptions used in the model (i.e., introduce additional factors)
4. Communicate the solution

The focus of much of traditional mathematics education, though, has been to emphasize one particular set of tools (algebraic manipulation) with little or no emphasis on any other tools or any other part of this process. However, it is certain that no one can anticipate the kinds of tools that will be available 10 years from now. Certainly, any routine procedure that is important has already been programmed and is available at the push of a button. If solving routine template problems continues to be a primary emphasis of mathematics courses, the education of our students will be outmoded before they leave the classroom. It is no longer defensible to offer courses in which the primary focus is having students practice one set of routine problems after another and then have them reproduce those solutions on an exam. In the past, one might have argued that this was unavoidable in order to develop the high algebraic skill level needed for calculus; today, that argument may no longer be valid.

8. RESISTANCE TO CHANGE

There are a variety of reasons for resistance to change in the precursor courses. Many are the same as those seen with calculus reform, although there are others that are unique to the courses below calculus. Consider the following challenges and possible solutions.

Challenge: For some, the resistance is purely philosophical. They feel that reform courses are changing the very nature of the subject. Many such individuals, particularly those who do not actually teach any of the courses, have a strong suspicion that reform is just a euphemism for watering down the courses.

Solution: This kind of resistance can often be overcome by showing the greater depth of mathematics being covered in the new courses and especially by demonstrating the deeper level of student understanding and achievement.

Challenge: For some, the key argument against reform is: "That worked for me and everyone else I know, so it is the only way to produce more of me."

Solution: The people who espouse such views are often new faculty members recently out of graduate school. They often just need a year or two in front of a classroom to learn quickly that the overwhelming majority of students are not interested in becoming mathematicians. In the past, there really was little in the way of an alternative to the traditional curriculum. However, with the options available today, a little support and guidance can often convince these faculty that curricular change can improve student learning, as well as the environment in their classes.

Challenge: Some fear, often with good reason, that something will be lost in the reform courses. If an instructor introduces new material or new emphases or has the students working in groups in class, it will be impossible to *cover* everything.

Solution: In the past, topics and methods once considered essential were eliminated from the curriculum—though at a much slower pace than is happening today. For example, look at what happened to the Law of Tangents.

On the other hand, it is reasonable to expect that a student who has learned some mathematics well at any given level of the curriculum and who has also learned how to think mathematically should be able to later pick up that textbook and independently learn something. For instance, if partial fractions are not taught in precalculus, a student with a solid precalculus-level background should be able to learn a little about partial fraction decompositions while taking Calculus II. Similarly, a student who has not been exposed to integration of rational functions via partial fractions in calculus should be able to read up on it quickly while taking a course in differential equations when using partial fractions to derive Laplace transforms. Of course, this freedom to change content within courses as appropriate requires open and frequent communication among instructors.

Challenge: Some mathematicians are cautious and uncomfortable with technology. They often have not fully thought through the positive implications for teaching mathematics. They may be concerned that students will ask questions that they are not prepared to answer, either about the technology itself or the different mathematical questions that can arise. Some may fear that the technology will replace the mathematics as the primary focus of the course.

Solution: Most such objections can be overcome by exposing these individuals to the technology in a supportive environment of faculty training workshops. However, the use of technology cannot be forced on instructors as something that must be used in class; their input and feedback must be part of the process of change.

Challenge: At many schools, the faculty teaching the precursor courses are people with relatively limited mathematical backgrounds and/or experience. They may be part-time faculty or graduate assistants, or individuals who have a very high comfort level with what they know and have been teaching for years, but do not have the personal confidence to try something new and different. Such individuals are often tentative about adopting renewal ideas because they lack knowledge about some of the mathematical subject matter. For instance, the notion of families of functions or slope fields or regression and correlation analysis may not have been part of their mathematical training.

Solution: Often, the most effective approach is to offer on-going training workshops, though local circumstances may well make it difficult to ensure that part-time faculty will be able to participate. Unfortunately, it is not always easy to secure the funds needed to pay for such participation. Nevertheless, it is worth the effort to explain to college administrators that mathematics is changing—and why—and request additional funding for faculty development. Evidence of improved student performance and retention, as well as closer ties with other quantitative subjects, is an important part of such a conversation.

Challenge: At some schools, the precursor courses are the responsibility of separate developmental departments and the faculty in these departments may not be aware of the changes that are taking place in calculus and other courses in the mathematics department. They likely will see their role as unchanging, even though the students in calculus are being held to a different standard and are having very different experiences in class, in their homework, and on their exams.

Solution: In such situations, there is an urgent need for on-going communication to acquaint these instructors with the new emphases in calculus and the changing skills that students now need.

Challenge: Many faculty, particularly at two-year institutions, carry very heavy teaching loads. In addition, reform courses are very labor and time intensive for the instructor. Nonroutine conceptual or applied problems can be difficult to construct; they are also more time-consuming to grade on exams or homework. Similarly, grading extensive written assignments and reports, as well as mentoring the student projects leading to such reports, is also time-consuming. It is little wonder that faculty are resistant to implementing reform courses, even when they agree philosophically with the goals of the courses.

Solution: A support system involving all members of a department can help make the transition easier. Often, a feeling of camaraderie can make the extra effort less burdensome and the opportunity to exchange ideas with colleagues can be a tremendous stimulus in developing new problems and examples. It is also helpful to include frequent reminders that the efforts are in the long-term best interests of the students. Finally, it may be possible to convince administrators to provide financial support and/or release time if a sufficiently strong case can be made that the extra time involved presents an overwhelming obstacle. If the rationale for introducing the new approach is compelling (say, the expectation of having a better success rate), it may well be a convincing argument.

Challenge: Faculty at two-year institutions are legitimately concerned about the transferability of reform courses.

Solution: According to *Assessing Calculus Reform Efforts,*[5] a significantly higher percentage of four year colleges and universities have been implementing calculus reform efforts than two year colleges. In particular, approximately 76% of all doctorate-granting institutions, about 74% of all master's-granting institutions, and 74% of all bachelor's-granting institutions reported having implemented either moderate- or large-scale calculus reform. In comparison, about 54% of two-year colleges reported such activities. Furthermore, almost all of the institutions that reported such efforts indicated that they anticipated larger-scale reform activities in subsequent years. Thus, it appears that transferability is likely not an issue at all. In fact, if an instructor at a two-year college does not teach in a reform style, the students may well be at a disadvantage when they transfer to a university that likely utilizes such techniques.

Challenge: One argument occasionally raised against reform is that students will not be adequately prepared for physics or chemistry or "X."

Solution: The changes that are taking place in the precursor courses, as well as in the reform calculus classes, very closely mirror comparable changes that are taking place in fields such as engineering, physics, and chemistry. These changes include a greater emphasis on conceptual understanding, a lessened emphasis on algebraic manipulation, group work on projects, the use of technology, and so forth (see Gordon[6]). While they may not be taking place in every department in every college, they are certainly happening with increasing frequency—especially in engineering departments that are accountable to the newly developed standards of the national Accreditation Board for Engineering and Technology (ABET). The faculty in other departments that are renewing their own courses are important colleagues with whom mathematicians should collaborate.

9. WHAT A MODERN PRECURSOR COURSE SHOULD EMPHASIZE: SUMMARY RECOMMENDATIONS

A good precursor course is one that takes a broader perspective than just preparing students for calculus, whether it is reform calculus or traditional calculus. The reality is that the overwhelming majority of the students will not go on to calculus and the precursor course must deliver something useful and important to everyone. The challenge is to provide a solid preparation for calculus while simultaneously meeting the long-term needs of all students. What then should such a course feature?

9.1. Students Must Understand Variables and Functions

- Students must see variables as representing the values of quantities and, in order to meet the needs of other disciplines, the letters used for variables should include more than just x and y.
- Students must achieve a solid understanding of the difference between independent and dependent variables.
- Students must see functions through multiple representations: as formulas, as graphs, as tables, as verbal depictions, and as dynamic processes that describe realistic phenomena; they must understand the perspective represented by each.

- Students must come to understand the practical limitations of any given function, in the sense of domain and range. In practice, domain and range mean far more than just looking for those points where one divides by zero or where one takes the square root of a negative number. For instance, if one creates a linear model for the world record times in the mile run based on data from the start of the twentieth century, it makes little sense to extend it backward to predict how fast Columbus or Plato could have run that distance.
- Students must recognize certain classes of functions both from formulas and from graphs; specifically, linear, exponential, power, polynomial, and sinusoidal functions. When a student sees a set of data, a chart, or a graph, he or she should be able to match it with functions that behave in the same manner.
- Students must understand fundamental ideas about the behavior of functions, such as growth versus decay, increasing versus decreasing, and concave up versus concave down.
- Students must understand the effect of various parameters on the behavior of functions. For instance, the slope of a line must be interpreted in the context of units of measure and scale, not just so many boxes over so many boxes.
- Students must understand the effects of transformations—stretching and shifting—on functions.

9.2. Applications Should Be a Unifying Theme for All Precursor Courses

The majority of students taking precursor courses are far more interested in the applications of mathematics than they are in the mathematics itself. Instructors should capitalize on this interest and, in the process of developing an appreciation for the power of mathematics, perhaps an appreciation for the beauty of the mathematics will follow. It is also possible to incorporate a tremendous wealth of mathematical thought through applications. In particular:

- Applications should be used to illustrate connections between the mathematics and the real world, connections with other courses, and (perhaps most importantly) connections to students' personal experiences.
- Applications—both the contexts and the mathematical models used—should be realistic instead of artificial expressions that are created only to have the answers work out "nicely."
- Applications should help students to see how real problems lead to equations. This realization should then be used to help students better

understand the meaning of a "solution."

- Applications should be used to help students explain mathematical ideas and solutions in everyday language, both oral and written. From a mathematical point of view, a solid explanation is the first step toward a formal proof.
- Students should be exposed to nonroutine problems that require applying mathematics to new situations.
- Applications of the mathematics should be used as motivation for students to take additional courses in mathematics.

9.3. Students Should See the "Big Picture"

Even the best students will take only a few ideas and methods with them when they leave a course. Instructors need to recognize this fact and make sure that students understand the fundamental concepts and methods of the course.

- Students should not be overwhelmed by too many specific examples at the expense of the main ideas.
- Students should see how certain fundamental ideas arise constantly throughout mathematics, not just during a unit on the topic. For example, exponential behavior is extremely common in real-world phenomena; linear behavior is certainly the most common. Therefore, exponential functions should not be approached as a topic to be covered in one or two weeks, with no connections to the rest of a course. Similarly, systems of linear equations should be seen as a unifying notion, not just a seemingly minor topic covered for a week or so.
- Students should spend less time on the mechanics of finding solutions for such equations and use this additional time learning to interpret such solutions.

9.4. Students Should Become Familiar with a Variety of Mathematical Tools—Symbolic, Graphical, and Numerical—and the Technology that Allows Us To Apply These Tools

Technology is everywhere and certainly will become far more prevalent in the future. The use of some type of technology, whether a graphing calculator, a spreadsheet, or a computer algebra package, should be an essential component of any modern precursor course.

Students should become comfortable with all of the different tools for solving equations. For example, any equation involving a single variable can be solved using graphical or numerical methods to any desired degree of accuracy;

certain particularly simple kinds of equations can be solved exactly using symbolic methods.

Rethinking the precursor courses to reflect these principles and implementing the necessary changes may well appear as a daunting challenge to many. However, the rewards—both for the students and for the instructor—are well worth the effort. As the reform movement continues to influence more of the mathematics curriculum, numerous textbooks and project materials are becoming available that embody these principles. Simultaneously, the needs of faculty preparing to teach from these texts are being addressed for the first time. Numerous faculty development workshops are being offered by professional organizations, by publishers, and by the individuals involved in the projects.

In addition, many written reports and professional presentations describe individual and departmentwide experiences with such courses. These reports provide guidance for implementing such programs and can help mathematicians and others understand that these changes to the precursor courses provide a unique opportunity to improve the mathematics education of students.

10. CONCLUDING REMARKS

There is an old Chinese curse that says: You should live in interesting times. Those in mathematics education likely feel that they have been hit with a multiple whammy of that curse. And, the reform movement is, if anything, merely gathering steam. Mathematicians will face increasing pressures for greater changes in the coming years.

Most of the changes discussed describe things that are happening at the college level, though they certainly have major implications for how secondary school students are prepared for college mathematics. At the same time, comparable changes are being implemented across the secondary curriculum as the NCTM *Standards* are more widely adopted and as traditional textbooks are replaced with more reform oriented textbooks. Thus, the colleges can expect to see increasing numbers of students with very different mathematical experiences and proficiency than previously. Students will have had less formal algebraic manipulation, but a greater emphasis on conceptual understanding and multiple representations of those concepts; consistent exposure to the use of technology, almost entirely in the form of graphing calculators; more experience in writing and communicating mathematics; more experience with extended, realistic applications of the mathematics; and more experience in learning in nontraditional classroom settings and with more varied methods of assessment used.

This, in turn, will put ever increasing pressures on those college faculty who have been resistant to curricular change in the hope that it will either fade away or that they can resist until retirement. As mentioned above, there are certainly many who have resisted on legitimate philosophical or other grounds. In the past, there has been a seamless transition between secondary and college mathematics because all schools (secondary and college level) offered the same courses in the same spirit. The reform efforts at all levels will soon create a very different— though equally seamless—transition between secondary school and college. But in the process, mathematicians should all expect a lot of bumps as the pieces rub against each other to generate the kind of perfect fit that they would like.

What also is clear is that the mathematics students coming out of this new curriculum model will be different. Instead of pencil and paper, these students will naturally turn first to some kind of graphical tool to investigate what a process looks like or whether an idea makes sense. They then will turn to the algebraic techniques only for general verification. Certainly, the next generation of mathematicians will not be clones of the current one; they will bring a very different vision to mathematics and related areas.

But what about the relatively few people—mathematicians, physicists, and computer scientists, for example—who will need to know the full array of symbolic operations? It now seems unreasonable to have everyone learn those skills, particularly as prerequisite skills to courses that may no longer require them. It is clear that this did not work well for the overwhelming majority of students, even when such skills were essential for success in calculus and related courses. Now that such skills are less necessary, there is no reason to expect all students to master them.

It may therefore be reasonable, in the not-too-distant future, to see junior- and senior-level college courses in advanced manipulative algebra offered at large research universities for the few students who really need it, while the overwhelming majority of students are exposed to more fundamental mathematical reasoning, mathematical ideas, and realistic applications.

It is evident that the secondary school mathematics plate has shifted dramatically and will continue to shift even further. The postsecondary mathematics plate may be shifting in the same directions, so that at some schools, there will continue to be a relatively smooth transition. However, many other schools may want to recall what occurs when the earth's plates shift in different directions; and an earthquake is not a pleasant experience.

REFERENCES

1. A. Solow (ed.), *Preparing for the New Calculus*, MAA Notes #36, (Mathematical Association of America, Washington, DC, 1994).
2. J. Dossey (ed.), *Confronting the Core Curriculum: Considering Change in the Undergraduate Mathematics Major*, MAA Notes #45 (Mathematical Association of America, Washington, DC, 1998).

3. Conference Board on the Mathematical Sciences, *Statistical Abstract of Undergraduate Programs in the Mathematical Sciences in the United States: Fall 1995 CBMS Survey*, D. O. Loftsgaarden, D. C. Rung, and A. E. Watkins (eds.), (Mathematical Association of America, Washington, DC, 1997).
4. *Curriculum and Evaluation Standards for School Mathematics*, (National Council of Teachers of Mathematics, Reston, VA, 1989).
5. A. Tucker and J. Leitzel, *Assessing Calculus Reform Efforts*, MAA Reports (Mathematical Association of America, Washington, DC, 1995).
6. S. P. Gordon, *A Roundtable Discussion with the Client Disciplines*, in W. Roberts (ed.), *Calculus: The Dynamics of Change*, MAA Notes #39 (Mathematical Association of America, Washington, DC, 1996) 152–156.

CHAPTER 7

Program Evaluation and Undergraduate Mathematics Renewal

The Impact of Calculus Reform on Student Performance in Subsequent Courses

Jack Bookman

Educational research and research with human subjects must, by its nature, be flawed—at least in comparison with the standards set by scientific research. Program evaluation, in particular, must deal with complex and difficult-to-control situations. Therefore, conclusions from such research must be tentative and qualified. Some argue that because of this complexity and ambiguity, program evaluation is not worth doing. However, program evaluation—flawed as it is—can provide valuable insight into what a program has accomplished and what components have contributed to or impeded its success.[1] This chapter will address a particular aspect of the evaluation of current calculus reform efforts, namely, the effect of calculus reform on student performance in subsequent courses. It will summarize the findings from several studies on this subject and will address the corresponding methodological difficulties.

1. SOME BACKGROUND ON CALCULUS REFORM

Between 1988 and 1997, the National Science Foundation, the primary source of financial support for the calculus reform movement, provided more than $40 million in various programs for the improvement of calculus instruction. The calculus reform projects have made changes in the content of what is taught in calculus classes, as well as (and perhaps even more fundamentally) changes in the way calculus is taught. The meaning of calculus reform varies greatly from one institution to another. However, reform is usually manifested by changes in one or more of the following: use of technology, cooperative learning, group projects, writing, and real world problems. Tucker and Leitzel[2] report that in a spring 1994 survey of 1048 postsecondary institutions, 22% reported undertaking major reform efforts while another 46% reported instituting modest reforms.

There has been a significant amount of work done to evaluate these efforts, though the quality of these studies varies immensely. In her report, *Ten Years of Calculus Reform*, Ganter[3] discusses the results of her exhaustive search of efforts to evaluate calculus reform. Of the 110 institutions that had one or more of the 127 NSF-funded calculus reform projects, about one-half reported conducting an evaluation study. These evaluations, along with others from programs not funded by NSF, addressed student achievement, student attitudes, student attrition, and faculty reactions. Though the results of these evaluations are mixed, Ganter reports the following general trends concerning student performance:

> ... evaluations conducted as part of the curriculum development projects revealed better conceptual understanding, higher retention rates, higher scores on common final exams, and higher confidence levels and involvement in mathematics for students in reform courses versus those in traditional courses; the effect on computational skills is uncertain.

> ... in general, regardless of the reform method used, the attitudes of students and faculty seem to be negative in the first year of implementation, with steady improvement in subsequent years if continuous revisions are made based on feedback.[3]

In her conclusions she also states that "the success or failure of a reform effort is not necessarily dependent upon what is implemented but rather how, by whom, and in what setting." This raises important issues for the design and interpretation of evaluation studies and points to just one of many factors that can complicate matters.

In her report, Ganter also states that very few studies exist on the long-term effects of calculus reform on student performance, i.e., studies that investigate student performance after calculus. In fact, since the calculus reform is just over 10 years old, reports of such studies are now beginning to surface.

2. SOME BACKGROUND ON PROGRAM EVALUATION

Popham[4] provides the following definition for systematic educational evaluation: "Systematic educational evaluation consists of a formal assessment of the worth of educational phenomena." In the 1960s, the federal government began to fund educational projects and required that these projects be formally evaluated. This demand for accountability, together with dissatisfaction with public education, stimulated the development of the field of educational evaluation. During this period, the scholarship of program evaluation began to develop, with Scriven and Stake among the most influential theorists.[4]

Scriven defined the distinction between formative and summative evaluation. Formative evaluation refers to evaluations of programs as they are being developed and still capable of being modified. Summative evaluations are statements of the worth of a completed instructional program.[4] In the words of Robert Stake, "When the cook tastes the soup, that's formative; when the guests taste the soup, that's summative."[1]

In planning evaluations, evaluators need to ask the following questions:

1. What are the goals of the program being evaluated and what is the purpose of the evaluation? In particular, what are the specific questions and the program that need to be answered?
2. What data need to be collected and how can that be done in an objective, unobtrusive fashion?
3. What techniques should be used to analyze the data?
4. To whom, and in what form, should the data, analysis, and conclusions be reported?

In answering these questions, other issues will arise that need to be addressed. For example, Who is the audience (or audiences) for the evaluation? Should the data collected be qualitative, quantitative, or some combination? Who will monitor the evaluation? A good evaluation, like any other research project, requires extensive planning, careful execution, and thorough reporting.

The implementation of the program evaluation must, by necessity, lag behind program development. When funding for program evaluation is part of the funding for program development, the evaluations often focus on formative and short-term summative questions, as has been the case with most of the evaluations of calculus renewal projects. Evaluators often feel required, or pressured, to complete their evaluation during the two- or three-year period for which the project development was funded. Longer term evaluation studies, such as those that will be described in this paper, require persistence and, sometimes, separate funding that will allow the evaluation research to continue after the development funding has ended.

3. THREE STUDIES OF THE IMPACT OF CALCULUS ON STUDENT PERFORMANCE IN SUBSEQUENT COURSES

Based on an exhaustive search by the author, the following three studies represent all available evaluations comparing the performance of students, from both traditional and reform calculus courses, in courses beyond calculus.

3.1. University of Illinois at Chicago

Beginning in 1995–96, the University of Illinois at Chicago changed its method of teaching Calculus I and II in their courses intended for science and engineering majors from a traditional course to a reform course.[5] This change includes the adoption of the Hughes-Hallett et al. textbook[6] and some changes in teaching style that include a workshop format in discussion sections in which the teaching assistant facilitates group work on specific problems. Calculus classes (both before and after the introduction of reform calculus) meet for three one-hour lectures with a faculty member and two one-hour recitation sessions with a graduate teaching assistant.

The goal of this study was to examine the grades of students (from either traditional or reform calculus) in courses having a calculus prerequisite, particularly in client disciplines such as physics and engineering. The subjects of the study included all students who began traditional calculus in 1994–95 (the last year that it was taught) and all students who began reform calculus in 1995–96 (the first year it was taught). There were about 1100 students in each of these groups and since the adoption was uniform across all sections, neither students nor faculty had a choice about whether to take part in the reform or traditional course, thereby eliminating the issue of self-selection.

For each of the two groups the following data were collected: Calculus I grade, method (traditional or reform), Mathematics-ACT score, Mathematics-Placement test grades, and the grades in subsequent mathematics, science, and engineering courses taken in the three semesters after they completed Calculus I. Using the Calculus I grade, Mathematics-ACT grade, and Mathematics Placement test grade as covariates, analysis of covariance was used. Statistically significant differences were found for grades in the following courses:

Favoring the reform course	Favoring the traditional course
Physics I	Physics II
C Programming I	Calculus III
Organic Chemistry I	
Calculus II	

The authors found the grade distribution in Physics I to be dramatically higher after reform and, in their paper, discuss these results in much detail. However, they report a significant decline in the performance of reform students when they take Physics II, but they don't explain this result. They note that Calculus III was and continues to be taught in a traditional manner and the differences observed are likely related to transition difficulties from reform calculus course to traditional calculus. The authors also report that "there is some indication that reform students get to higher courses in larger numbers over the same period—14 reform to 4 traditional in the courses observed."[5]

3.2. University of Arizona

The University of Arizona was one of the original members of the Harvard Calculus Consortium, teaching reform calculus for the first time during the 1990–91 academic year. Beginning in 1993–94, all first-year calculus courses were taught using the Hughes-Hallett *et al.* textbook.[6] Alexander[7] examined university grade records to determine if there was a difference between reform and traditional students in retention and grades in subsequent calculus-dependent courses. He defines the reform course as one using the Hughes-Hallett *et al.* textbook[6] and the traditional course as one using a traditional text. He describes the difference in the content of the two courses but does not mention differences in pedagogy.

Six cohorts of students were identified for which comparison data existed. Since both reform and traditional calculus was taught during the 1991–92 and 1992–93 academic years, the experiences of students in Calculus I (Fall 91, Spring 92, Fall 92) and Calculus II (Spring 92, Fall 92, Spring 93) were examined. Many client courses were originally considered for the study, but only four courses could be identified for which there were significant numbers on which to base a comparison:

Calculus III	Ordinary Differential Equations
Physics I	Statics (a Civil Engineering course)

For each of the cohorts in each of these client courses, the grade point average (both including and excluding students who withdrew from the course) and retention rates (the ratio of the number of students who attempted the client course to the number who passed the applicable Calculus course) were computed. Separate retention rates and grade point averages were then computed for those students who took the course at the normal time and for those who eventually took the course. Therefore, for each cohort and each course, six different measures were computed.

The results were mixed. For each of the courses, between 25 and 33% of the 36 measures produced significant differences with the bulk (31 out of 43) in favor of reform calculus. As Alexander points out, many of the differences occurred with the first group of students in the study who were taught by a volunteer group of enthusiastic instructors. However, when observing only the next group (the students in Fall 92 Calculus I and Spring 93 Calculus II, where Alexander reports there were fewer instructor differences), 11 of those favored the reform students—even though there were only 13 of 48 measures with significant differences. Clearly, this is not overwhelming evidence for the effectiveness of the reform course, but the data indicate that there may be some real differences that further study could clarify. In summarizing his results he states, "The evidence for improved performance by consortium students is weak, however there is good evidence that consortium students are not more poorly prepared than are their peers in traditional courses."[7]

3.3. Duke University

As part of a larger evaluation study[8-10] of Project CALC, Duke University's calculus reform project, a study of student performance in courses following calculus was conducted. Project CALC differs from the traditional Calculus course in several fundamental ways.[11,12] The traditional course emphasizes acquisition of computational skills, whereas the key features of Project CALC are real-world problems, activities and explorations, writing, teamwork, and use of technology.

The follow-up study had two sets of goals: to examine the attitudes of students two years after completing freshman calculus and to examine their performance in mathematics-related courses. More specifically, answers were sought to the following questions:

1. In general, how well did students do in the mathematics and science courses they took after having a year of college calculus—both traditional and reform?
2. In particular, how well did they do in mathematics courses? How well did they do in engineering courses?
3. Did reform students take more or fewer mathematics courses during their college careers?
4. What factors may have contributed to these differences?
5. What aspects of their experience in freshman calculus do students attribute as being important in their experiences in subsequent courses?

During the 1990–91 academic year, approximately 29% of the students enrolled in first-year calculus were randomly assigned to Project CALC (PC) sections; the remaining students were assigned to traditional (TR) sections. However, per university policy, students were allowed to drop and add courses during the first two weeks of classes. As will be addressed later, this may have had a significant impact on the results of the study. The subjects of the study were students who had completed a year of TR or a year of PC. For each subject in the study (78 PC students and 182 TR students) the following information was collected in January 1993: SAT–mathematics score, SAT–verbal score, major, grades in Calculus I and II, overall GPA at Duke, and the students' grades in the following courses:

- Any Physics course
- Any Engineering course
- Any Economics course
- Any Computer Science course
- Any Mathematics course (beyond the level of Calculus III)
- Introductory Biology (usually taken after one or two semesters of calculus)
- Organic Chemistry
- Public Policy Analysis

Except for the courses in Economics or Public Policy Analysis, these are all science, mathematics, engineering, and technology (SMET) courses. Promoting success in SMET courses has been one of the goals of calculus renewal, so it was important to include these courses in this study. Economics and Public Policy Analysis are also courses requiring quantitative thinking, so they were included as well. An average mathematics-related grade (MSGR) was computed for each student who had completed at least two of the listed courses. To remove distinctions between more and less leniently graded courses, MSGR was computed by averaging the difference between the subject's grade in a course from the average for that course. It was found that on average, the grades of the traditionally taught students were better by 0.2 of a 4-point GPA. This is a small but statistically significant difference. It remained statistically significant even when SAT scores and performance in Calculus I and II were controlled.

In order to better understand these results, seven pairs (one PC, one TR) of these students—matched by major and SAT scores—were interviewed. The purpose of the interviews was to find out: (1) if the subjects felt adequately prepared for their future courses; (2) if their attitudes about mathematics had changed as a result of their experience here; and (3) how, in retrospect, they felt about their experience in mathematics at Duke. Most of the students felt that they were adequately prepared for future courses, but some students, particularly the engineers who were veterans of Project CALC, expressed disappointment that the use of computing technology in later courses was limited. They reported that this was especially true for their subsequent engineering courses. In general, the PC students were less confident in their pencil-and-paper computational skills, but more confident that they understood how calculus and mathematics in general are used to solve real-world problems.[13]

In January 1994, the grades of the same group of subjects (now seniors) were examined again. The results of the analysis of the composite grades were similar to those of the previous year: there was about a 0.2 (out of 4.0) difference in favor of the TR students that remained statistically significant even after controlling for SAT scores. Also examined were subsets of the data consisting only of grades in those SMET courses that had significant mathematical content—courses in engineering, physics, computer science, and mathematics. In addition, a separate analysis was conducted on just the grades in mathematics courses. In each of these cases, there was virtually no difference between the average grades of PC and TR students. PC students took a larger number (though not statistically significantly so) of these more mathematical SMET classes and, in particular, significantly more courses offered by the mathematics department. Twenty of eighty-seven PC students took more than two mathematics courses beyond Calculus III as opposed to 27 out of 207 for the TR group. The difference (23% versus 13%) was statistically significant ($p < 0.05$). PC students also performed slightly better in those courses than their TR peers. There was also virtually no difference in the grades of the PC and TR engineering students in engineering courses. Table 1 summarizes these results. Retention rates for both groups from Calculus I to Calculus II were consistent with the historical totals.

It seems that although there were few differences between the grades of the two groups in SMET courses, the significant differences overall may be explained by differences in the performance in the other classes observed, most notably Introductory Biology (see Table 2). A possible explanation for the poorer performance of this group of PC students in Biology may be related to the fact that students were not strictly randomly assigned (they were free to change classes during the two-week drop–add period). It is hypothesized that a significant number of the more conservative, more risk-aversive students (who at Duke University are disproportionately premedical students) may have transferred out

Table 1. Performance in SMET Courses with Significant Mathematical Content

	PC			TR			
	n	Mean	SD	n	Mean	SD	p
Overall grades							
In mathematical SMET courses	55	2.90	0.62	130	2.89	0.69	0.91
Mathematical grades	20	2.66	0.79	27	2.57	0.70	0.68
Engineering grades	23	3.07	0.72	45	3.13	0.68	0.74
Average number of	87	5.68	6.77	207	4.77	6.19	0.28
mathematical SMET courses							

Table 2. Performance in Introductory Biology, Organic Chemistry, and Economics

	PC			TR			
	n	Mean	SD	n	Mean	SD	p
Biology 1	36	2.196	0.830	87	2.766	0.754	0.0001
Biology 2	37	2.246	0.696	86	2.676	0.718	0.002
Organic Chem 1	37	2.341	0.792	92	2.433	1.013	0.631
Organic Chem 2	29	2.359	0.922	68	2.649	0.948	0.16
Economics courses	11	2.809	0.470	40	2.928	0.735	0.516

of the PC course and into the TR course. If true, this has major implications for the possible interpretations of the results of this study.

4. CONCLUSIONS AND RECOMMENDATIONS

These studies indicate that although calculus renewal is no magic pill, it may have some positive effect on both the performance and continuation rates of students in SMET classes. The evidence is not uniform across all courses nor is it overwhelming or dramatic, but each of the separately conducted studies produced similar results. Despite some occasional anecdotal evidence to the contrary, there appears to be no systematically collected evidence to show that calculus renewal has hindered academic performance in SMET courses. However, three isolated studies in an enterprise as vast as calculus education in the United States are far from conclusive evidence; many more such studies need to be conducted to determine the validity of these conclusions with any degree of certainty.

These project evaluations were designed to assess the extent to which the various projects at the individual institutions studied were successful. They were designed to answer the following questions: (1) How well did students— traditional and reform—do in the mathematics and science courses after having a year of college calculus? (2) Did students from reform calculus courses take more or fewer such courses during their college careers than traditionally taught students? The studies were conducted in an objective manner consistent with the standards of program evaluation research. However, these studies did not (and did not intend to) test any learning theories or explain, in a detailed or theoretical manner, why reform calculus seems to have some positive effect on student performance in SMET courses.

Learning theory and cognitive psychology experiments have traditionally been conducted in controlled laboratory environments involving relatively small numbers of subjects. The goal in such research is often to test a particular hypothesis about how learning occurs and to examine learning in a careful and microscopic way. What happens when the results from this setting are applied to real and complex situations involving lots of subjects in impossible-to-control environments? What happens when we "scale up" from the laboratory environment to the classroom and institution, i.e., when we move from controlled environments to uncontrolled (and often out of control) environments? These situations—messy, complex, and fraught with methodological potholes—are the environment in which program evaluation must be done.

One of the most difficult issues—and the issue that is perhaps most problematic in interpreting the results of the evaluation studies described above—is the inability to randomly assign students to control and experimental groups. After randomly assigning students to either reform or traditional sections, students are often free to change sections during a drop–add period. This is a situation that is usually unavoidable; the whole university cannot be converted into a controlled learning environment. However, it is possible to track those students who switch sections in order to better understand the extent to which this may eventually affect results.

In medical research, double-blind studies (where subjects to be studied are unaware whether they have been given a placebo or the treatment and the examiner is unaware whether the subject has been given a placebo or the treatment) are the standard. This standard is certainly not practically applicable here. And although it is helpful, it is not enough to simply control for a particular set of measurable variables such as SAT scores. It is difficult—and perhaps not possible—to control for or measure (in a valid and reliable way) such traits as willingness to take intellectual risks. It is also difficult—and probably impossible—to know a priori what traits should be controlled, even if we could measure them. This does not mean that one should not conduct such studies, only that (1) methods of assigning students to groups must be carefully reported and (2) these difficulties in interpreting the results must be considered. The data don't "say it all"; they need to be interpreted. In particular, evaluators need to explain how and whether the data collected help answer the questions asked and whether the data can be interpreted in more than one way, particularly in light of sampling problems.

One way to address some of these problems is by coupling qualitative and quantitative research methods. Qualitative data, including interviews, observations, and examining documents, can provide valuable insight into the interpretation of quantitative data.[14] Quantitative data can, in turn, support the evidence and add objectivity to the results from qualitative data. For example, interviews with students two years after they completed calculus revealed that former Project CALC students reported benefiting from the increased emphasis on conceptual understanding, practice in explaining, and applying mathematics to real-world problems; traditional students did not report benefits from the corresponding mathematical formalism of their course.[13] This could explain the increase in the number of subsequent mathematics courses taken by reform students.

Although grades in courses are convenient and readily available, they are not a very reliable (and, arguably, not a valid) measure of student learning. In the University of Illinois study described, the authors note that Calculus III was and continues to be taught in a traditional manner and that the differences in grades (favoring traditionally taught students) likely reflect the difficulty in transition from a reform calculus course to a traditional course. Similarly, some of the students interviewed at Duke reported difficulty in making a transition to traditional courses. One student reported, concerning his experience in linear algebra, that "after coming out of Project CALC, I was so used to thinking in terms of application, in terms of where is this going to be used, and now here I am, memorizing different matrix operations, which I had no idea of why I was doing and consequently (and I am not ashamed of this) I got a C in the class."[8] Evaluation studies conducted in the future need to develop other instruments for assessing what students have learned that will be more reliable, valid, and universally accepted than course grades.

There is still great debate in the mathematics community about the effectiveness of calculus renewal, and there is a great need for careful, objective, and thoughtful evaluation to inform this debate. In particular, there is a need for more studies (like those described in this chapter) that examine the long-term effects of calculus renewal on student achievement, attitudes about mathematics, and careers. In addition, research needs to be conducted on describing (after the initial enthusiasm and external funding for reform have subsided) what aspects of calculus renewal have the most enduring impact on how calculus is taught. This research should address questions such as:

- Is the mainstream calculus today fundamentally different than 10 years ago, or just superficially so?
- Which ideas of calculus renewal are enduring and which are fleeting? Which should be maintained?

Although more evaluations concerning the effect of calculus reform need to be conducted, it should not be expected that the results of such work will lead to some mathematical truth about how calculus is learned. However, we can hope that a reasonable person would be persuaded by the preponderance of evidence collected and that the accumulated results of these studies do, in fact, paint a realistic picture of the effect that such projects have on students' understanding of calculus.

REFERENCES

1. F. Stevens, F. Lawrenz, and L. Sharp, *User-Friendly Handbook for Project Evaluation Science, Mathematics, Engineering and Technology Education* (National Science Foundation, Washington, DC, 1993).

2. A. C. Tucker and J. R. C. Leitzel, *Assessing Calculus Reform Efforts* (Mathematical Association of America, Washington, DC, 1995).

3. S. L. Ganter, *Ten Years of Calculus Reform: A report on evaluation efforts and national impact* (Mathematical Association of America, Washington, DC, in press).

4. W. J. Popham, *Educational Evaluation* (Prentice–Hall, Englewood Cliffs, NJ, 1975).

5. J. L. Baxter, D. Majumdar, and S. D. Smith, "Subsequent-Grades Assessment of Traditional and Reform Calculus," *PRIMUS* (in press).

6. D. Hughes-Hallett, A. Gleason, *et al.*, *Calculus* (Wiley, New York, 1994).

7. E. H. Alexander, "An Investigation of the Results of a Change in Calculus Instruction at the University of Arizona," 1997, unpublished doctoral dissertation.

8. J. Bookman and C. P. Friedman, "Final Report: Evaluation of Project CALC 1989–1993; 1994, unpublished manuscript.

9. J. Bookman and C. P. Friedman, *The Evaluation of Project CALC at Duke University 1989–1994.* (Mathematical Association of America's Notes Series, in press).

10. J. Bookman and C. P. Friedman, in E. Dubinsky, A. H. Schoenfeld, and J. Kaput (eds.), *Research in Collegiate Mathematics Education I* (American Mathematical Society, Providence, 1994) 101–116.

11. D. A. Smith and L. C. Moore, in T. W. Tucker (ed.), *Priming the Calculus Pump: Innovations and Resources* (Mathematical Association of America, Washington, DC, 1990) 51–74.

12. D. A. Smith and L. C. Moore, in L. C. Leinbach (ed.), *The Laboratory Approach to Teaching Calculus*, (Mathematical Association of America, Washington, DC, 1990) 81–92.

13. J. Bookman and C. P. Friedman, "Student Attitudes and Calculus Reform," *School Science and Mathematics* (March 1998):117–122.

14. E. G. Guba and Y. S. Lincoln, *Effective Evaluation* (Jossey–Bass, San Francisco, 1981).

CHAPTER 8

Redesigning the Calculus Sequence at a Research University

Faculty, Professional Development, and Institutional Issues

Harvey B. Keynes, Andrea M. Olson, Dan O'Loughlin, and Douglas Shaw

1. INTRODUCTION

One of the major objectives of many reformed calculus projects is to create a challenging sequence that helps students better understand how to use calculus as a tool both for mathematical analysis and for solving problems in other disciplines. This objective is widely recognized by the scientific community.

> Although most of the mathematical research community is university-based, the impact of mathematics on society is pervasive. Mathematics underpins most current scientific and technological activities. The applications of mathematics in the future will require closer partnerships between mathematical scientists and the broader universe of scientists and engineers.... Hence, the mathematical sciences are now essential to all three aspects of science: observation, theory, and simulation.[1]

Implementation of this philosophy requires changes in curriculum, content, methods of instruction, student–faculty, mathematics faculty, and interdepartmental faculty interactions. The curriculum needs to be initially developed by interdisciplinary faculty teams. Then the mathematics instructional teams, which include faculty, postdoctoral students, teaching specialists, graduate and sometimes undergraduate students, need to discuss and plan specific curriculum and pedagogical directions and actively share in the ongoing development for each course in the sequence. Overall, a calculus sequence that attempts to address the issues raised by the scientific community requires a major reconceptualization and reorganization of the traditional large lecture/recitation approach used in many doctoral institutions, and must incorporate key aspects of calculus reform efforts enunciated by the mathematics community.

> At the 1986 Tulane Conference, professors of various client disciplines as well as a diverse group of mathematicians met to evaluate traditional calculus approaches and to propose effective strategies for accomplishing change. They concluded that the traditional calculus sequence did not meet the needs of the students for subsequent math courses or those of client disciplines whose foundation is built on mathematics. They agreed that students would obtain a greater conceptual understanding via applications and numerical explorations with computers and graphing calculators by using traditional algebraic methods as a guiding theme.[2]

However, as noted in the 1995 Mathematical Association of America (MAA) report, *Assessing Calculus Reform Efforts*, very few doctoral institutions critically use projects or writing assignments or place major emphasis on modeling and applications.[2] Also, careful evaluation of student achievement and feedback from client disciplines have been only minimally addressed (Tucker and Leitzel,[2] p. 33), and student feedback during the course is frequently ignored. Emphasizing these and other features typically minimized at doctoral institutions (creative approaches to encouraging student-centered pedagogy, increasing personal interaction between faculty and students, and supporting the effective use of various technologies) presents challenges for faculty and administrators who wish to successfully implement these aspects.

Most reformed calculus courses want to go beyond rigorous paper-and-pencil computation and allow students to use appropriate technologies to pose mathematical problems based on scientific phenomena. In addition, students may be required to professionally write their solutions. "As an example, in one application, students are asked why a rainbow always seems to hit the horizon at the same angle—about 42 degrees—and why the color distribution is the same."[3] Traditionalists in the mathematics community believe that these approaches are innovative but view them with skepticism. Will students be able to perform complicated mathematical calculations if they become dependent on graphing calculators and computers? One critic of calculus reform, George E. Andrews, head of the mathematics department at Pennsylvania State University, states,

"The reformers believe that they will get around the roadblocks of basic arithmetic so students can get to higher-order skills."[3] To address such issues, several reform models incorporated routine calculations into gateway exams to insist that all students completing the course achieve necessary computational skills. Even though students are often engaged in group work and other methods of active learning, the University of Minnesota Calculus Initiative (CI) clearly found that there are certain situations where the lecture method is the best form of pedagogy. This observation agrees with the findings of University of California, Berkeley mathematician H. Wu who states that "lectures are very effective, and in fact there are even circumstances which make this method of instruction mandatory."[4] In fact, the overall central role of the faculty lecturers was strengthened by the CIs structure at the University of Minnesota and was highly evaluated by students. Overall, when the curricular and pedagogical elements suggested by the scientific community become integral components of calculus coursework, they appear to have a positive impact on learning.

The Institute of Technology Center for Educational Programs (ITCEP) and School of Mathematics, University of Minnesota, developed a new reformed calculus sequence for science, engineering, and mathematics students. The primary objective of the CI, piloted in 1995–96 and institutionalized in 1996–97, was to enable Institute of Technology (IT) undergraduates to better learn calculus and the critical thinking skills necessary to apply it in a variety of science and engineering disciplines. The reconceptualization of the traditional approach to calculus developed by the Initiative has integrated both content and instructional changes that emphasize active learning through the use of teamwork and increased personal interaction between students and faculty, and has emphasized applications in addition to computational skills and supporting technologies. These approaches were designed to improve the perception of the culture of mathematics at a large research institution and, more importantly, influence student understanding of how to use calculus as a tool for mathematics and problem solving across the disciplines.

This paper will discuss how the model developed for this project requires instructional and policy decision changes to effectively implement a reformed calculus sequence at a large research university. Future implications for similar sequences at other institutions, based on the progress, successes, and challenges of the University of Minnesota's initiative, are also discussed.

2. PHILOSOPHY, OBJECTIVES, AND ORGANIZATION OF THE CALCULUS INITIATIVE

Student objectives for the CI were opportunities for active learning, creative uses of lecturing and other pedagogical methods, increased student/faculty contact, and increased exposure to the conceptual and visual aspects of calculus. To achieve these student goals, instructional goals for the CI instructors were to increase the: (1) range/depth/quality of the material taught at each level, (2) professional satisfaction of each instructor, (3) personal interactions between students and faculty that would more effectively engage students in the subject, and most importantly, (4) student learning and retention of the concepts and methods of calculus. In order to meet these objectives, major shifts in faculty interaction and professional development were implemented.

Two underlying policies integral to the project were accountability and sustainability. Ongoing reviews of the project's impact on the faculty and the policies surrounding these issues have been conducted. In addition, specific criteria and evaluative processes were established to measure student attitudes about the usefulness of the pedagogy, how this approach affected their learning of calculus, and the impact the CI had on the students' problem-solving skills in subsequent courses. Continuing assessment of the CI's effectiveness is a key aspect to maintaining its quality.

The CI features student-centered learning involving group work and appropriate technologies. This is in contrast to the general audience calculus sequence that uses the traditional lecture/recitation approach with a more standard curriculum. Any calculus-ready (using a standard mathematics placement test), nonhonors science, engineering, or mathematics student is eligible to enroll in the CI. Over the past four years, University of Minnesota student achievement and retention in the sequence has been above average. (University of Minnesota students are not permitted to continue in the sequence if they receive a course grade of less than C−.)

The mathematical content of the CI is fairly traditional. In the first year (Y1) of the sequence, single variable calculus is covered completely and differential equations and multivariable calculus are introduced. The second year (Y2) covers multivariable calculus through Gauss/Green/Stokes and differential equations. One new aspect is an introduction to numerical analysis with an emphasis on computer lab applications taught at the end of Y2.

The CI is structured to allow a wide variety of useful pedagogical techniques to be incorporated, while still being manageable and cost-effective. Its six main components are listed in Table 1.

Table 2 summarizes the fall term student/staff ratio for Y1 and Y2.

It is important to note two aspects of Table 2. First, despite enrollment expansion, the amount of administrative effort to manage the CI continues to decrease. This results from ongoing efforts to systematize the materials and procedures into a format that permits undergraduate student workers and part-time coordinators to handle most administrative duties. Second, the lack of interested and pedagogically qualified mathematics graduate students has led to the increased use of teaching specialists (primarily outstanding secondary

Table 1. Six Main Components of the CI

1. Two 50-minute weekly class sessions of 100 students that primarily use a lecture format but also include up to 15 minutes of group work per session.
2. Two workshop sessions per week of 25 students, one 100-minute session and one 50-minute session, consisting primarily of group work (in Y1) or computer labs (in Y2), which encourage active learning and peer collaboration. Some homework is discussed and short quizzes are given in the workshops.
3. Periodic visits to the workshops by the lecturer, and visits by the workshop leaders to the lectures, help facilitate the group work.
4. Three large-scale team projects (some include computer lab applications) are assigned in Y1. In Y2, weekly computer labs using appropriate applications take place. Half of the labs require a written report, and one lab is assigned over several weeks as a group project.
5. In Y1, "gateway exams" on differentiation and integration consist of standard computations to be completed at a high level without the use of a calculator.
6. Texts with rich applications, i.e., *Calculus from Graphical, Numerical, and Symbolic Points of View* by Ostebee and Zorn or *Calculus: Concepts and Contexts (both SV and MV)* by J. Stewart, *Differential Equations* by Blanchard, Devaney, and Hall, and *Vector Calculus* by Barr, are used in both Y1 and Y2.

Table 2. Composition of the CI Staff

1995–96 Y1-1	1996–97 Y1-2	1997–98 Y1-3	1998–99 Y1-4
100 students	200 students	300 students	400 students
2 co-lecturers	2 lecturers	3 lecturers	4 lecturers
1 workshop administrator (Postdoc)	2 co-workshop administrators (graduate students)	1 workshop administrator (postdoc)	1 workshop administrator (postdoc)
4 workshop leaders (2 postdocs, 2 graduate students)	8 workshop leaders (graduate students)	12 workshop leaders (3 postdocs, 3 graduate students, 6 teaching specialists)	16 workshop leaders (3 postdocs, 4 graduate students, 1 undergraduate, 8 teaching specialists)
	Y2-1	Y2-2	Y2-3
	75 students	120 students	220 students
	1 lecturer	2 lecturers	2 lecturers
	1 workshop administrator (graduate student)	1 workshop administrator (postdoc)	1 workshop administrator (teaching specialist)
	3 workshop leaders (graduate students)	5 workshop leaders (graduate students)	9 workshop leaders (1 postdoc, 5 graduate students, 1 undergraduate, 2 teaching specialists)
1 sr. administrator (33%)	1 sr. administrator (25%)	1 sr. administrator (15%)	1 sr. administrator (10%)

teachers). Moreover, as of the fall of 1998, both the first and second year of the sequence are piloting the use of several outstanding undergraduate upper-division students as workshop leaders. The lecturers and workshop administrators are closely mentoring the undergraduate teaching assistants (TAs) and a complete assessment of this pilot will be done at the end of the academic year.

Reducing the administrative effort necessary to manage a sequence and expanding the potential pool of qualified workshop instructors/TAs appear to be critical components in maintaining the quality of a reformed calculus sequence. These changes influence policy decisions by requiring increased efforts for training/professional development for support staff and instructional team members.

3. ADMINISTRATIVE AND INSTRUCTIONAL FLEXIBILITY WITH THE SEQUENCE STRUCTURE

3.1. Administrative Efficiencies

The initial administrative efforts to establish the criteria for sequence eligibility have been greatly reduced by basing eligibility on the student's standard calculus placement exam scores, which are taken at freshman orientation. Over the past four years these scores have proven to be a very good predictor of student performance. Other administrative efforts associated with the CI are the production and distribution of the workshop and lab materials, maintenance of the labs, tracking of student achievement and retention, and ongoing assessment of student and faculty outcomes. Effectively managing part-time and undergraduate student employees to handle the majority of the responsibilities can minimize the cost for these efforts. This management requires cooperation and close collaboration with the instructional teams, and most critically, the course chair and workshop administrator.

3.2. Collaborative (Team) Approach

The structure of the CI was deliberately designed to encourage the use of various pedagogical techniques in both the large and small group sessions. Drill, practice, and conceptual discovery are integrated into both lecture and workshop, as opposed to reserving lecture for concepts and workshop for drills. Moreover, materials are carefully linked to topics frequently foreshadowed in one venue and then formally taught in the other. This necessitates a good deal of uniformity in the workshops and lectures, which is partially addressed by having the activities designed by the senior faculty and workshop administrator. In addition, successful implementation requires increased student–faculty, mathematics faculty, and interdepartmental faculty interactions. Another important aspect is the improvement of student attitudes about mathematics, and the use and value of mathematical reasoning in other disciplines. Students need to be involved in the analysis of problems that interlace mathematical concepts with important applications while working as part of a student team in a workshop setting. The teams are expected to collect and analyze data, to formulate and test conjectures, and to communicate their ideas clearly and effectively through oral and written reports. The combined use of experimental, visual, numerical, and analytical approaches in the modules develops skills critical for such analysis in a variety of disciplines, and is increasingly used in mathematics. Moreover, students are required to become more actively involved in the learning process and meet the increased responsibility for acquiring routine computational skills.

A "stratified" approach is stressed for each applications-based project and laboratory module involving the analysis of science- and engineering-based problems. The introductory layers of the material stress the fundamental mathematical concepts while the more probing layers allow students to explore topics in greater detail and help make connections with previous (and future) topics. All layers foster teamwork and stress the importance of clear communication of mathematical ideas.

A feature critical to maintaining the performance and motivation of all faculty and workshop leaders is the discussion and planning of specific curriculum and pedagogical directions, and their active involvement in its ongoing development. Through electronic communications and regular weekly meetings, all members have the opportunity to contribute to the discussions, make suggestions for restructuring, and continue the dialogue on how the course is progressing. In addition, ongoing close contact between the instructional team and the students allows rapid course changes in response to student feedback. Changes in homework due dates, weekly workload distribution, and weight of course components were implemented as a direct result of team discussions.

Criticism of the usefulness of the team approach is often based on the assumption that the value added is outweighed by the increased cost for additional instructional staff. However, the CI effectively minimizes the need for additional resources over the traditional large lecture model. Small classes of 25 students during three of the five hours each week create excellent opportunities for small-group activities and experiences, and provide the overall atmosphere of a small classroom with modest additional instructional costs. Other cost-effective measures have been the use of teaching specialists, specifically outstanding secondary teachers as workshop leaders and outstanding undergraduates in lieu of graduate teaching assistants. Personnel specifically hired for teaching are compensated for their instructional abilities and not for any other research or service activities.

3.3. The Influence of the Secondary Teachers

In the 1997–98 academic year, retired or sabbatical secondary teachers who had successfully taught high school calculus were workshop leaders for 5 of the 12 Y1 sections of the CI at Minnesota. Students' exam scores and final grades were generally higher, with differences narrowing as the material became more advanced. These teachers generally had the highest overall rankings on student evaluations, and their office-hour attendance tended to be higher, probably based on their extensive skills and experience in working individually with students. The students seemed to be more comfortable interacting with instructors similar to their secondary teachers. Also, the teachers' enthusiasm and dedication to students probably increased student effort. These instructors voluntarily conducted tutorial sessions prior to major exams and held extended office hours (sometimes on weekends) to better meet their students' needs.

The secondary teachers provided valuable input throughout Y1. One secondary teacher was assigned to visit and mentor the class of a graduate student TA having difficulty facilitating group work. His students showed marked improvement. Certain single-variable workshop activities were modified based on the secondary teachers' input.

The secondary teachers rated the professional development aspects of their CI experiences very highly. They reported that in addition to learning mathematics, they were able to incorporate the techniques and materials of the Initiative into their own high school courses. They very much enjoyed many of the conveniences, mathematical focus, and instructor respect that teachers in a college setting take for granted. All of them reported that they would recommend this experience to their colleagues as an excellent professional development experience, and more teachers are asking to become involved. In the 1998–99 academic year, 8 of the 16 workshop sections are being taught by secondary teachers.

There were some procedural changes that were necessary to accommodate the background of the secondary teachers. The teachers insisted on more time to look through the workshop materials, and requested solutions for all of the group work. Several requested more frequent meetings with the course chair and the workshop administrator to discuss both the subject matter and the pedagogy of the course. Overall, the secondary teachers' influence on the Initiative was very positive.

3.4. Undergraduate TAs

The involvement of selected undergraduate TAs was piloted in 1998–99. Most of these students were graduates of the CI classes and are currently outstanding junior/senior students. With careful professional development and ongoing mentoring, these TAs can be successful instructors. However, there are several issues to be addressed with the practice of hiring undergraduate TAs. For example, during exam periods, can the undergraduates balance this heavy responsibility with their own course load? When students tend to rely most on their TAs the undergraduate TAs may be occupied both with their own examinations and with job interviews. Can these TAs handle the full range of content issues involved in a quality calculus course? Will it become difficult to hire exceptional undergraduates as the course enrollment increases? Can quality mentoring programs still be maintained as the number of undergraduate TAs increase?

To ensure a successful experience for secondary teachers and undergraduate TAs and to maintain quality of instruction, longer preparatory times, complete instructors' solutions guides, and increased professional development are essential. The primary benefit of increasing the availability of nonfaculty instructors to assist the faculty lecturers without increasing instructional costs is very significant.

3.5. The Instructional Team Approach and Its Impact on Students

In order to have a unified and successful approach among the instructional team members, a consistently visible and supportive environment is essential. This is created by having the lecturers visit the workshops and the workshop leaders assist in the lectures. The visitations help to create a built-in mentoring aspect. Professors, postdoctoral students, undergraduate and graduate students, and teaching specialists need to be encouraged to create an atmosphere of collaboration and commitment to quality. Discussions about mathematics in addition to specific course content and pedagogy are generated at the weekly instructional team meetings. These elements allow the lecturers and workshop leaders to succeed within the structure of the CI and to improve as instructors. Professional development is a natural result of using the team approach; it has been particularly observable in instructors who were involved in the CI at Minnesota for more than one year.

This approach also increases student interaction with faculty and workshop instructors. It personalizes a traditionally large, multisection course; faculty get to know students' names, students become more approachable, and likewise, faculty are viewed as approachable by the students. This interaction also provides a direct connection between the lecture and the workshops. Student response to material in the workshop helps to shape the content, approach, and delivery of the next day's lecture material to better fit the students' current level of understanding. The faculty are able to present a broader perspective of the material in the lecture— linking it to previously studied concepts and workshop material, covering new material and foreshadowing upcoming theory.

The results of the students' behavior who participated in the CI at the University of Minnesota indicate they noticed the instructional team's consistency and enthusiasm, and that it had a positive influence on them. Students in the CI were surprisingly willing to do more work than their peers in the traditional course. For example, a sizable number of students who were not required to enroll in the entire sequence did so anyway. During evaluations, the students felt free to give constructive criticism and believed that it would be seriously considered. Although attendance was encouraged, it was not required at lectures or at a majority of the workshops. However, attendance in the workshops and, even more surprising, in the lectures was extremely high (an average of 90% across three years). A senior University of Minnesota faculty lecturer remarked that, overall, students in this course were more focused and eager to learn than they were in the standard courses.

At most large research universities, instructors and TAs will have a range of teaching experiences, and not all of them will be outstanding instructors. The result of using a carefully constructed instructional team approach is greater uniformity in successful outcomes over the different sections than is typical in a large multisectioned course. Throughout the Minnesota experience, the variance in common exam averages between the top and bottom sections was smaller than expected. One lecturer had a teaching style that was substantially less popular than the others. Rather than not attend the lectures at all, a fair number of his students sat in on the other lectures and continued to attend their regular workshops. Virtually none dropped this course. Most importantly, the final grades of the students in this lecture section were nearly identical to the other lecture sections and enrollment retention in the sequence was not affected.

3.6. Use of Technology

The technology used to support learning in this model varies as the sequence progresses. In the first year students extensively use graphing calculators with a few forays into modules that utilize interactive programs over the Internet. During the second year, students use graphing calculators and appropriate computer applications and symbolic algebra systems in regular laboratory settings. Ideally, the essential component of these materials should be made accessible to classrooms where technology is limited to graphing calculators. For example, an initial optics lab application written for the CI at Minnesota was rewritten so that it required only a graphing calculator.

The large-scale, computer-based projects or applications, which allow students to use the mathematics they have learned, take place during specific workshop/lab sessions. Teams of three or four students work during a few scheduled sessions in which they run the experiments and simulations, ask questions, and get help as needed with expectations that significant time spent out of class is essential. Availability over the World Wide Web allows the student teams, in principle, to work the entire lab at many locations. Each team develops a lab report over the next few weeks in which 10 to 20 extended questions are addressed. Students are required to allocate labor, oversee quality, and make design decisions as well as learn mathematics. In short, they practice being part of a research team and this process allows students to work cooperatively while increasing individual skills and knowledge.

3.7. Summary of the Quantitative Outcomes Analysis: A Retrospective Survey

In addition to surveys conducted during the academic year in the CI courses, a retrospective survey was developed to investigate student attitudes about the CI and its value in upper-division IT coursework. Several attempts were made to contact the 78 students who had completed *at least* the second quarter of Y1-1— by telephone, surface mail, and e-mail survey. A very robust response rate of 64% (50/78 students) was obtained. Table 3 lists the demographic data and the objective and subjective survey components.

In the combined analysis of all 50 surveys, 7 of the 11 categories that were similar to the midquarter evaluations were highly rated by at least 80%. For the components that compared CI attributes to other mathematics, science, and engineering courses, 5 of the 8 items were rated "better" by at least 80%. Personal contact with faculty and instructor/TA attention to learning achieved a standard of at least 50% of the students giving the highest rating. Clearly, students continued to especially value interaction with the instructional team members.

When separately analyzing the results of the female students, an even more positive picture emerges. Fourteen of the nineteen categories were highly rated by 80% of the students, and 55% gave 10 of 19 categories top ratings. Thus, the female cohort seemed to be the most satisfied group with the overall CI structure and outcomes. It is interesting to note that this cohort included many students who left IT to become accounting, business, and history majors.

The overall student responses to the subjective questions mirror the objective responses. In evaluating the two most important components, the most frequent responses were "personal contact with instructors/TAs, instructor/TA core for students, group work, and technology." Thirty-three of forty-seven students positively rated the role of the CI as preparation for other science, mathematics, and engineering courses, and only two expressed concerns. Thirty of forty-nine students responded that the Initiative influenced their interest in mathematics, while only two indicated a decrease. Thirty-five students indicated that the CI increased their confidence to use mathematics, while only four were negative. Many students expressed positive comments about the CI overall. For example, one student wrote: "I wish all IT departments would participate in such programs

Table 3. Retrospective Survey

Demographics of the 50 Survey Respondents
   25 Telephone Survey Respondents
      Gender: 18 male, 7 female
      Year in college: 21 seniors, 3 juniors, 1 sophomore
      Current major: 10 engineering, 8 math/statistics/computer science, 3 other sciences,
      1 accounting, 1 business, 1 history, 1 undecided
      10 completed the first year of the Initiative
      15 completed 2/3 quarters of second year of the Initiative
   25 surface mail/e-mail survey respondents
      Gender: 21 male, 4 female
      Year in college: 23 seniors, 2 juniors
      Current major: 11 engineering, 6 math/statistics/computer science, 5 other sciences,
                          1 architecture, 1 political science, 1 communication disorders
    1 completed first year of the Initiative
    24 completed 2/3 quarters of second year of the Initiative
Objective components
   Rate the usefulness of the following components using a scale of 1 to 5
   (1 = not useful, 2 = limited use, 3 = moderately useful, 4 = useful, 5 = extremely useful):
      a. Small class size
      b. Personal contact with instructor
      c. Office hours
      d. Workshops
      e. Personal contact with workshop leader
      f. Small group work
      g. Collaboration with peers
      h. Graphing calculators
      i. Homework
      j. Exams
      k. Computer labs
   Rate the aspects listed below comparing the CI to other IT courses using a scale of 1 to
   5 (1 = CI was worse, 2 = CI was slightly worse, 3 = CI was about the same, 4 = CI was slightly
   better, 5 = CI was much better):
      a. Learning environment
      b. Instructor/TA attention to learning
      c. Lectures
      d. Workshops
      e. Computer labs
      f. Coordination of lecture, workshop, and lab
      g. Encouragement of collaboration with peers
      h. Instructor/TA accessibility
Subjective components
- If you were given the choice now to take the Standard Calculus Sequence or the Calculus Initiative, which would you choose?
- In your opinion what were the one or two most important components of your Calculus Initiative courses in terms of helping you learn calculus?
- In your opinion what one or two aspects of the courses would you change in order to enhance your learning of calculus?
- What was the influence of your Calculus Initiative courses on your other science, mathematics, and engineering courses?
- Did your Calculus Initiative courses increase or decrease your interest in taking other math courses? Explain how.
- Did your Calculus Initiative courses affect your confidence in your ability to use calculus effectively in other courses? If yes, explain.
- Please give any other comments you feel would be useful to this study.

for helping students learn, gain confidence in themselves, and make friends the way the mathematics department did in creating the Calculus Initiative."

In summary, the results of the retrospective surveys indicated that one to two years after completing the sequence the students remained extremely positive about the value and impact of the CI on their learning, preparation for other coursework, and interest in using mathematics throughout their collegiate career. All students, virtually without exception, indicated that they would retake the CI if offered the choice at this time. Thus, the Initiative clearly achieved its goal of helping students learn how to better use mathematics in related Institute of Technology coursework.

Most faculty members and graduate students felt that their effectiveness as teachers improved as a result of teaching the sequence. Several postdoctoral students, graduate students, and senior faculty who taught in the CI had questions about modifying traditional calculus and some concerns about the course structure. In the end, virtually all of them valued their experience and eliminated their personal doubts about the teaching approaches. Some actively recommended teaching in the CI to other colleagues and others volunteered to continue to teach in this sequence (believing that student learning was improved).

The structure of the CI clearly provided opportunities for instructor development. The instructional teams had opportunities for interaction in both the lectures and the workshops. Their weekly meetings often centered on teaching and content issues as well as administrative aspects. The result was improvement in overall teaching approaches and more satisfaction from teaching in the CI. One international graduate student TA, who was not considered to be a particularly effective instructor, matured to the point that after teaching two quarters of the CI she was a much more confident and successful teacher and her students were specifically requesting to continue in her section. The instructional development aspect of the CI is becoming even more evident as the pilot program using undergraduate students as workshop instructors is evolving.

4. IMPLICATIONS OF SUCH MODELS FOR LARGE RESEARCH UNIVERSITIES

4.1. Long-Term Prospects for the Course

The Initiative has generally enjoyed support and respect from the mathematics, science, and engineering faculty at the University. Initial achievements have brought the CI to the position where currently enough sections have been opened to guarantee that every qualified student in the engineering college who wishes to take the CI is permitted to do so (96% made this election in 1998–99).

There are two major issues that could affect the future of the Initiative, despite its success. The first is cost. Even after implementing all reasonable economies, there is an incremental cost difference of 20–25% over the standard calculus sequence to offer the current CI model. This unavoidable cost differential is mainly attributable to the extra workshop instructor and administrative efforts necessary to support its more personalized structure. Despite the achievements and successes of the approach, colleges and departments will need to balance other priorities with provisions for any incremental funding. For example, while the School of Mathematics continues to actively support the CI and seek additional internal funding, the long-term outcome remains uncertain. Perhaps the enthusiasm shown by the Institute of Technology Dean and most science/engineering departments will have a positive effect on its future viability. The authors of this paper believe that should the CI be required to operate with the same resources as the standard calculus sequence, the CI model would most likely seriously compromise some of its best features. Hopefully, the increased use of teaching specialists and the anticipated use of outstanding upper-division undergraduates as workshop instructors will continue to help reduce costs. But consistent departmental and institutional financial and intellectual support are critical to maintaining the quality and viability of the CI model of calculus.

The second issue is staffing. Changes in graduate student interest and availability and the overall need for ongoing professional development are critical issues for large research institutions. The identification and training processes of the TAs require additional time and resources. The current CI model relies on an ongoing cadre of key faculty and postdocs to provide the teaching and content leadership as well as the TA training. Continued interest, availability, and professional motivation of a group of faculty to address these aspects are central to maintaining the quality of the CI. As professional development becomes more of a concern across all disciplines, there may be ways to integrate the continued development necessary for the Initiative instructors with other departments and colleges. This would help to reduce fiscal and human resources, as well as promoting collaborative effects throughout the university. Nevertheless, departmental and institutional rewards and recognition for both graduate students and faculty need to be consistently supported to encourage involvement of talented staff.

4.2. Continuing Need for Professional Development

In a recently released national report, *Transforming Undergraduate Education in Science, Mathematics, Engineering, and Technology,*[7] one key recommendation was for more extensive opportunities for the faculty to learn best teaching practices and techniques to improve instruction. A parallel outcome was to provide all graduate students with opportunities to improve teaching skills and learn innovative practices. These types of efforts in an even more extended fashion are key to maintaining a quality CI model.

Professional development and mentoring as addressed in the previous section for faculty and graduate students directly associated with the CI is also needed for beginning undergraduate student TAs, especially in the areas of small group work, development of workshop projects, and effective use of technologies. Also, postdoctoral students and some senior faculty need some professional development in how to use active learning methods during lectures and how to effectively mentor TAs in group activities. As the Initiative continues to grow, more formal professional development will be needed to encourage new under-graduate TAs, graduate TAs, and faculty to participate in the CI as well as to update current staff. One already identified critical component will be professional training in the effective use of technology in a reformed calculus model.

4.3. The Role of Technology

The least sustainable element of the CI model is the computer labs in the second year. Most universities have a limited number of computer labs available compared with student need, and most institutions require that labs must be shared with other department or college courses. Based on the existing curricular model (Y2-3) of the second-year CI at the university, major computer labs were required for 25–30 hours per week so as facilitate the student technology requirement. In addition, the Y2-3 lecturers spend an average of 12 hours per week helping the students and assisting the workshop instructors in the labs. As the course enrollment increases, it will not be possible to proportionately increase these resources. Moreover, it is unrealistic to expect a large number of even interested faculty to devote the time and energy to this aspect without more resources and more compelling evidence of the positive influence of technology as student learning. In fact, current use of technology has raised questions and inhibited some faculty from further involvement in the CI. Other large research universities will have to face similar issues.

Transferring some of the computer applications to more advanced calculators (such as the TI-89) is being investigated as one way to address some of the issues addressed above. Hopefully, a sufficiently advanced calculator could be used for some aspects of the assignments currently requiring a computer, and new assignments more compatible with a calculator could be developed. Initially, it would be cost-effective to loan this type of calculator (appropriately programmed) to the students during lab time in lieu of purchasing additional computer equipment and setting up new labs. Ultimately, as the cost of the advanced calculators stabilizes, the students (who are currently required to purchase a graphing calculator for the sequence) would be purchasing these advanced models for the course. Initial piloting of a TI-89 calculator project assignment produced quite positive student evaluations and a great deal of interest in purchasing such a calculator for its mathematical benefits.

Several new approaches to distributed learning may also help to support the lab component of the CI. One of the most promising approaches, currently being explored at the University of Minnesota, is CD-ROM applets that provide workshop-type interactive explorations of important concepts and extensions of areas and problems in calculus. Some sample topics have been developed and the initial reactions have been positive. Other aspects to be considered include supplementary coverage of some topics with additional written materials (for the textbooks and the CD-ROM applets) and various types of communication and testing management systems. It is likely that some of these approaches will be piloted in the CI courses within the next few years at the University. To the extent that they contribute to learning, and reduce faculty efforts and real costs, these distributed learning approaches will be considered for incorporation into the CI.

5. SUMMARY AND RECOMMENDATIONS

The major influences on the future of the CI model are the instructional and policy decisions that will be made for this challenging calculus sequence for science, engineering, and mathematics students. Preliminary analysis indicates that a positive relationship exists between certain policy directions and decisions and improved instructional and learning outcomes. Sufficient fiscal support and ongoing academic leadership are key elements to sustaining the quality of such initiatives at large research universities. Two other critical factors that contribute to the success of such initiatives are (1) ongoing faculty interest and motivation and (2) administrative skill and experience, which take responsibility for professional development and training that includes appropriate uses of technologies and distributed learning approaches, in addition to curriculum review and development.

Successful models allow for growth and adaptation to changing institutional needs and pressures as well as student academic backgrounds, intellect, interests, and career motivation. Faculty focusing narrowly on just curricular and content objectives without attention to providing a more engaging environment in which students learn mathematics both for its intrinsic interest as well as for its applications has traditionally produced less successful student outcomes than the efforts of programs such as the CI. While such initiatives require more faculty and administrative engagement and leadership, the Minnesota model has certainly exemplified the value-added to student learning, faculty development, and the culture of mathematics by these extra efforts.

Management of both the academic and administrative components requires different levels of cooperation and coordination to ensure successful ongoing efforts. These include faculty and other instructional staff acceptance and adherence to a core set of instructional and administrative procedures and practices without significant deviation. For example, weekly team meetings among faculty and other instructors have proven to be powerful tools for keeping the course on a desired path, and it is worth devoting the hour required each week. Also, regular faculty attendance at workshop sessions is such a powerful message to the workshop leaders and especially the students about the role and interest of the faculty that the entire procedure is compromised when this aspect is removed. The team effort turns out to be greater than the sum of the individual efforts, and is the glue that makes the course outcomes so much more cohesive and improved.

This chapter has presented a framework for modifying an engineering/physical sciences calculus sequence, emphasizing the challenges of successful implementation and sustainability at a large research university. Of course, one of the major issues in assessing calculus reform or any type of educational activity is whether the successes/achievements of the effort align with the beliefs and value of the community. So it is probably useful at this point to place this framework into the context of a set of goals and outcomes for calculus renewal efforts that address many of the important components suggested by the greater engineering and scientific community.

David Smith[8] presents an interesting model for calculus renewal in Chapter 3 of this volume. Using some ideas from the psychology of learning and cognitive sciences found in *Applying the Seven Principles of Good Practice in Under-graduate Education*,[9] and supported by recent experimentation, Smith first considers the type of attributes that *employers* of university graduates have consistently rated as most important. One conclusion is that to the extent that a calculus course both teaches the desired content and encourages the use of these teaching principles, it probably is doing a better job of both helping students learn the materials and developing desirable attributes necessary for future coursework and careers. "Meeting the complexity of tomorrow's challenges will demand insights across the full spectrum of the mathematical sciences. Both the theoretical and the industrial impact of this development will be enormous."[2]

The authors of this paper believe that the instructional, professional development, and learning framework developed for the CI has substantially addressed both types of the best teaching practices and student outcomes cited above. There is certainly room for improvement as the model continues to evolve. However, through the continued use of the major components of the CI, many universities can improve the learning environment and the ability of its science and engineering students to use calculus.

Acknowledgments

We would like to especially thank Michelle D. Bell, Texas Southern University, Terry J. Williams, Lincoln University, and Jennie Nash, University of Minnesota, who conducted the retrospective student telephone surveys and helped compile the data while working as summer interns at the University of Minnesota.

The development of the Calculus Initiative was supported in part by a grant from the National Science Foundation, and the initial computer lab applications were developed with the support of the NSF-Geometry Center.

REFERENCES

1. Committee on Science, Engineering and Public Policy (COSEPUP), *The Mathematical Sciences: Their Structure and Contributions. Report of the Senior Assessment Panel of the International Assessment of the U.S. Mathematical Sciences (GPRA Report)* (National Science Foundation, Arlington, VA, 1998) 8–14.
2. A. Tucker and J. James (eds.) *Assessing Calculus Reform Efforts*, (Mathematical Association of America, Washington, DC, 1995).
3. R. Wilson, "A Decade of Teaching Reform Calculus Has Been a Disaster, Critics Charge," *The Chronicle of Higher Education* 43(Feb. 7, 1997):A12–A13.
4. H. Wu, *The Joy of Lecturing—With a Critique of the Romantic Tradition in Education Writing*, Department of Mathematics #3840. (University of California. Berkeley, 1998) 1–12.
5. H. Keynes and A. Olson, *Redesigning the Calculus Sequence at a Research University: Issues, Implementation, and Objectives*, Proceedings of the International Commission on Mathematics Instruction (ICMI), Singapore, 1998.
6. H. Keynes and A. Olson, *Calculus Reform as a Lever for Changing Curriculum and Instruction*, Proceedings Fourth World Conference on Engineering Education, St. Paul, MN, 1995, pp. 248–251.
7. Committee on Undergraduate Science Education Center for Science, Mathematics, and Engineering Education National Research Council (NRC), *Transforming Undergraduate Education in Science, Mathematics, Engineering, and Technology* (National Academy Press, Washington, DC, 1999).
8. D. Smith, "Renewal in Collegiate Mathematics Education," *Documenta Mathematica* Extra Volume ICM (1998):777–786.
9. A. W. Chickering and Z. F. Gamson (eds.), "Applying the Seven Principles of Good Practice in Undergraduate Education," *in New Directions for Teaching and Learning* No. 47 (Jossey–Bass, San Francisco, 1991).

CHAPTER 9

Politics and Professional Beliefs in Evaluation

The Case of Calculus Renewal

Alphonse Buccino

1. INTRODUCTION

Policy is a big industry in Washington, D.C.—making it, criticizing it, or changing it. Policy is often like a cloud of galactic dust. Sometimes it coalesces into a star, but mostly it remains a cloud of dust whose particles are sometimes discernible. A policy wonk is someone who tries to discern policy or articulate what it is even when there isn't any. *Webster's Dictionary* defines "wonk" as a student who studies excessively, which helps understand what "policy wonk" really means. There are a lot more policy wonks than policymakers, but it is not always easy to tell them apart.

Evaluation of federal agency programs explicitly or implicitly serves policy and policymakers are a significant audience. But policy wonks also get involved in funding evaluations or interpolating evaluations for policymakers. This chapter reviews some of the features of evaluation in this context. The argument is that evaluation has to be considered in relation to the way government programs originate and are designed and implemented. Consequently, the problems and issues of one are necessarily tied to those of the other. So the title of this chapter, refers to evaluation in its political context, sometimes referred to as policy-making. Two policy streams affect our story: science policy and education policy. Moreover, belief systems such as those House[1] calls *ideology* and *saga*, get tangled up with policy and politics to affect evaluation, particularly in the case of calculus reform. House's exceptional paper came to my attention late in the preparation of this chapter. Realizing that it is now 25 years old, it is surprising and dismaying that the impact of the paper has been far less than it merits. The effects of politics and belief systems on evaluation need to be addressed much more robustly in evaluation design and practice.

2. PURE SCIENCE: AUTONOMOUS OR PUBLIC?

For science and science education programs, the seeds were planted in 1945 for the fruits of policy we live with today. The post-World War II creation of science policy grew out of the debate on the plan proposed in the report *Science— The Endless Frontier*.[2] Science autonomy was the big issue and it played itself out in terms of elitism versus populism, nicely examined by Kevles,[3] and the relative implications of the tension between them for research and education. The scientific establishment, led by Vannevar Bush, the author of the report, argued strenuously for formation of a government agency that would be responsible to scientists through a part-time board of scientists. A bill to establish a National Science Foundation (NSF) passed the Congress and came to President Truman in 1947 with a recommendation from the Bureau of the Budget that it be vetoed on the following grounds[4]:

> The provisions of this bill regarding the vesting of part-time officials with full administrative and political responsibility, the virtual nullification of the President's appointment power, and the interference with the President's authority to coordinate and correlate governmental programs, are at such variance with established notions of responsible government in a democracy, that the President should withhold his approval.

President Truman did veto the bill in 1947. It was two years before a new bill came forward from Congress, this time keeping the National Science Board (the part-time officials referred to above), but making the director of NSF a presidential appointee rather than an appointee of the board. In retrospect, it seems incredible that the science establishment should have fought so hard for a point they were bound to lose in the end. The issue of autonomy of a government agency against the principal of executive responsibility and accountability may have seemed arcane then, and was perhaps not foreseen or understood until it came up in the end in the Bureau of the Budget's technical analysis. But once it is brought up, the Budget Bureau's view seems quite formidable. It might well be that this history suggests why many people, including policymakers, think of

scientists as *just another interest group*[5] rather than according science the special status that its practitioners would like it to have.

The politics of the distribution of NSF funds and the seeds for programs like the Experimental Program to Stimulate Competitive Research (EPSCORE), perennial issues at NSF, were also present at the political creation. For example, an amendment that failed in the 1947 debate was a provision that there be a geographical distribution formula in the allocation of NSF funding of basic research. When this amendment was defeated, Senator Richard Russell of Georgia asserted[6]:

> I unhesitatingly put in the Record the prediction, here and now, that within six years after the pending bill shall have been enacted into law, in the absence of an amendment of this nature, the two institutions referred to [MIT and Harvard] will be receiving more funds than all the educational institutions in at least the 12 states east of the Mississippi River and south of the Potomac.

The delay from 1945 to 1950 was costly to the science establishment in terms of its role and status in government. During that period, the Cold War emerged and heated up. This factor was instrumental in dividing responsibility for basic research across several agencies including military agencies like the Office of Naval Research rather than concentrate it in a super science agency envisioned in the Bush report. Nevertheless, the founding of NSF was the effective beginning of explicit support of government for basic research. Although the provision regarding distribution of funds was not included in the legislation, populists like Senator Russell and Senator Warren Magnuson of Washington did see potential in a connection they saw between research and education to produce scientists.

3. SCIENCE RESEARCH OR SCIENCE EDUCATION?

A major landmark in the politics of elitism versus populism in science was reached on January 30, 1956, the day the Subcommittee on Independent Offices of the House Appropriations Committee held a hearing on the NSF's request for fiscal year 1957. This hearing was momentous and fully captured by Kreighbaum and Rawson.[7] NSF had a small program of Teacher Institutes the agency called an "experiment." The discussions between members of the subcommittee and NSF spokespersons were confused "in the rapid cross-fire of questions and answers" without clarity of concepts, goals, or definitions. They included research and education programs at NSF and their relative priority. The hearing ended with Congressman Joe L. Evins of Tennessee asking the NSF Director Alan Waterman:

> Could you use $10 million instead of the three million dollars [in NSF's budget request] for high school teacher-training programs at this time?

Mr. Waterman was asked by the committee chair to "send us a little note on this."

In a letter four days later, Mr. Waterman, on behalf of NSF, declined the offer of $10 million for teacher institutes. The reasons given in the letter for turning down the offer were quite general, but the real reasons very likely included genuine practical reservations about NSF's ability to ramp up the program so rapidly and maintain the kind of quality NSF strove for and for which it had become noted in government and academe. But it also reflected the agency's priority to provide support for basic research and graduate fellowships above elementary and secondary teacher education.

Congressional action on the NSF budget that year resulted in a total appropriation of about $40 million for all NSF programs, nearly $6 million less than originally requested by NSF. It also included an extraordinary provision, later called a "limitation clause," that not less than $9.5 million shall be available for "... supplementary training for high school science and mathematics teachers", that is, for teacher institutes. This meant support for teacher institutes would have to be increased despite an overall reduction in the NSF budget request, effectively forcing a shift in NSF priorities from support for basic research to education. That same limitation clause (although the amount changed from year to year) remained in every NSF appropriation from fiscal years 1957 to 1971. It was clear that the Congress would have a role in setting broad NSF priorities and this has continued to the present.

4. EDUCATION POLICY: EXCELLENCE OR EQUITY?

A second policy stream parallel to science policy supported the continuing growth in the 1960s of the federal role in education in science, mathematics, engineering, and technology (SMET) at NSF and other agencies, and other aspects of education in the Department of Education, the National Endowment for the Humanities (NEH), and the National Endowment for the Arts (NEA). It received its impetus from the Great Society programs and the War on Poverty of the Johnson administration.

Education policy was derived from *human capital*, an area of economic scholarship that goes back to the origins of economics as a scholarly discipline. Adam Smith recognized human capital as "all of the acquired and useful abilities of all the inhabitants of a country" and urged that they be considered as a part of capital. While economists have always recognized human capital, it is frequently discarded as unrealistic because human beings are not marketable. In other words, human capital has not been seen as accessible to economic analyses. So while there is broad agreement that human capital and its cultivation is significant, there is not very strong agreement about the role of the federal government in human capital areas like health and education. This policy dilemma is quite evident in today's political environment.

Nevertheless, the economist Theodore Schultz[8] stimulated renewed interest in human capital among economists in his Presidential Address to the American Society of Economists. Significant increase of economic scholarship in human capital followed. A colleague of Schultz, Gary Becker, has published an important compendium of scholarship in three editions, 1964, 1975, and 1993, respectively[9]. The Nobel Prize in Economics has been awarded to Becker and Schultz separately for their respective work. These ideas strongly influenced policy that brought forth the Great Society programs and the War on Poverty of the Johnson administration. Among other programs, the federal support for education, heretofore seen largely as the responsibility of the states, increased dramatically through the then Office of Education (which became the Department of Education in 1980). Education funding through the Office of Education dwarfed that at NSF and still does. For example, federal revenues for public elementary and secondary education in the United States (NCES, 1996) grew from $355 million in 1953–54 (or 4.5% of total revenues) to about $2 billion in 1965–66 (or 7.9% of total revenues) and was about $18 billion in 1993–94 (or 7.0% of total revenues).[10] In comparison, NSF total support for all levels of education peaked at $120 million in the late 1960s, began a decline to a low of $15 million in 1984, and then began to grow to nearly $700 million in 1999. However, the ratio of education support to total NSF funding (including research) is lower today than it was at the post-Sputnik peak in 1969.

5. GROWTH OF THE RESEARCH AND EDUCATION ENTERPRISE

NSF also played a role in the Great Society programs. The emphasis was on an idea pushed by President Johnson for the spread of excellence by the buildup of science capability in every state. Programs were established at NSF to fund institutional and departmental development grants to increase research capability. The 1960s saw an enormous growth of both the research and education enterprises. Elementary and secondary education doubled in size (from about 20 million to about 40 million enrollments) in the decade between 1955 and 1965 and then leveled off. This, of course, reflects population growth—the postwar baby boom. But higher education in the United States not only doubled in enrollments between 1955 and 1965, but also doubled again between 1965 and 1975 as an increasing proportion of the age cohort attended college. Access to education was a great success in the United States as the nation brought a larger number of its citizens, in both absolute and relative terms, to higher levels of education. Also, the high-school dropout rate, which had declined slowly throughout most of the twentieth century, continued to do so until about the mid-1960s.

6. BENIGN NEGLECT OF EDUCATION

But by 1969, when Richard Nixon was inaugurated, the policy environment was overwhelmed with issues relating to the Vietnam War. While not an overt policy, budget problems again caused a reexamination of NSF priorities. This time, pressure to support applied research led to the establishment of the Research Applied to National Needs (RANN) program at NSF, a move away from an exclusive focus on basic research. There was also pressure to reduce support for science education programs. Given that education at NSF was to focus on increasing and strengthening the science and engineering work force, there were arguments that this need had diminished significantly by 1970 from what it had been in 1957. Moreover, the achievement of the goal to reach the moon (i.e., the success of the Apollo program) led to a decline of overall government support for research and development. In this time of tightening budgets, NSF and the White House Office of Management and Budget (successor of the Bureau of the Budget) once again wanted to give higher priority to research than to education.

There was a standoff with Congress that always liked the populist and distributed features of NSF science education funding, so the science education programs did not disappear. However, they were reduced and refocused on the newly emerged issue of groups underrepresented in the science and engineering work force; specifically, women and minorities. "Benign neglect" was one of the terms used at the time to describe the policy impact on human services programs during the Nixon administration.

In this environment, a curious battle arose in 1975 regarding NSFs education program. The focal point of the attack was *Man: A Course of Study* (MACOS), a curriculum materials development project for a fifth-grade course in social science. After spending 6 million NSF dollars, commercial publishers were uninterested in publishing the materials because they thought there wasn't an adequate market in sight. So NSF also paid for the publication of the materials through a small, private firm. At the same time, NSF teacher education and curriculum development programs were tied together more closely into an Implementation Program so that there would be greater efficiency in helping schools implement NSF-supported curricula (and other curricula as well).[11]

The attack on MACOS and NSF occurred in a climate when Congress expressed concern about agency (including NSF) funding of projects with titles that made them seem trivial to some observers. The projects in question were research into issues like insect mating calls, romantic love, and chimpanzee behavior. Easy to ridicule, Senator Proxmire (D-WI) initiated his *Fleece of the Month Award* even though there was always a robust rationale for supporting the grants so identified.

In any case, the initial assault on MACOS and NSF was led by the conservative John Conlon (R-AZ). He objected strenuously to this use of federal money to support a school curriculum that espoused situational ethics for impressionable fifth-graders. The innocent funding for publication was portrayed as part of a dark plot to brainwash U.S. children. Moreover, the growth of the Implementation Program, which was actually a response to earlier Congressional pressure, was portrayed by Mr. Conlon as "promotion and marketing" of the objectionable MACOS program. So a whole cluster of unrelated phenomena, quite benign and carefully following constraints and directions provided by Congress and the Office of Management and Budget, became tied together as a dark plot to undermine the nation's values and morality.

It was quite a commotion, partly because Congress is slow to reign in any of its members. One long-term consequence that emerged was an absolute prohibition on federal support of curriculum development, which particularly impacted the U.S. Department of Education as well as NSF. It also had a chilling effect. Was it advantageous to support science education when it could cause so much trouble? This kind of problem certainly has an impact on the program and evaluations of them.

7. EXTINCTION OF SCIENCE EDUCATION AT NSF AND ITS REAPPEARANCE

The Reagan administration came to Washington strongly opposed to federal support for education. The administration's opposition was focused on an absence of a federal role, not on the intrinsic value and merit of education per se. Parsing this kind of distinction is something many policy wonks like to do, but policymakers worry about it only when their respective ox is being gored or they want to gore that of another policymaker. At the same time, the specter of Soviet prowess came to the fore again, which led to the perfectly clear policy response of a big buildup in funding for research, particularly for the military, which included the Strategic Defense Initiative (sometimes called Star Wars) and a reduction for education. Specifically, science education at NSF was nearly eliminated entirely by the Reagan administration in 1983. All science education programs were terminated except the program of graduate fellowships, which continued at a level of $15 million per year.

But 1983 was also the year that the prestigious National Commission on Excellence in Education, appointed by the Secretary of Education, Terrell Bell, published its report *A Nation at Risk: The Imperative for Educational Reform*.[12]

(The current use of the term *reform* in connection with education dates from this report.) The Reagan White House did not leap to provide added funding for educational reform, but exhorted the states as the focal point of responsibility for education in the United States to undertake needed actions (seen as a form of federal devolution at the time and revisited by the Republican-controlled 104th Congress). Many states did take action, like tightening up on graduation requirements, but few were able to increase funding. One of them was South Carolina, which passed a 1% increase in state sales tax for education use exclusively. At the time, South Carolina's governor was Richard W. Riley, which is undoubtedly a factor of his being appointed Secretary of Education in the Clinton administration.

However, the reaction in Congress was somewhat different. Although science education at NSF was on the brink of extinction, a request came from Capitol Hill in FY1984, asking what NSF might do with an additional $15 million for science education. This was reminiscent of the "Could you use $10 million" question of 1956. NSF was as wary in 1984 as it was in 1956.

The response from NSF, which had to pass through the White House Office of Management and Budget, was a cautious program. Mindful of the MACOS hassle and prohibition on curriculum development, it proposed a modest program focused on elementary and secondary teacher education. Turnabout came from an unexpected quarter of Congress. The conservative Jake Garn (R-UT) wrote a letter of high rhetorical style, stinging NSF for submitting such a mild and cautious plan and then called for a heroic plan of great expanse that included a heavy emphasis on curriculum development but still, for the moment, within the $15 million limit earlier provided. Nevertheless, this change of heart about science education at NSF by the conservative Senator Garn, led to major growth in education funding at NSF over the next few years that matched the major growth in research funding initiated a few years earlier. Education grew from nearly zero to nearly $700 million annually, a level reached in FY1999.

Another irony is that about the time the heroic letter of Jake Garn arrived at NSF, the General Accounting Office (GAO), the investigative arm of Congress, published a report on programs for teacher education. Its summary finding[13]:

> GAO does not find evidence that training programs to upgrade existing science and mathematics teachers will improve teaching effectiveness. Such programs are a prominent part of proposed federal legislation.

The proposed federal legislation referred to was, of course, the same plan to which Jake Garn had responded in strong manner and tone. Evidently, GAO was not impressed with "training programs to upgrade existing science and mathematics teachers". Nevertheless, these programs have been, without doubt, the most politically popular of all science education programs from the time in 1956 when Representative Joe Evins (D-TN) singled them out for special treatment as something good to support without reference to outcome measures or evaluation.

8. TODAY'S BLUEPRINT: PATHWAYS TO EXCELLENCE

So despite the resistance of the Reagan White House to education, Congress once again began building education programs at NSF. In due course, George Bush ascended to the Presidency and called the historic meeting of the nation's governors in 1989, marking another significant milestone for U.S. education policy. The National Education Goals were adopted by the President and the nation's governors in 1990 to be met by the year 2000. Moreover, a national program called America 2000 was undertaken to implement them.

It is interesting to note that Goal 4 asserts that by the year 2000, "U.S. students will be first in the world in science and mathematics." This goal attracted a great deal of attention, especially among education professionals. Is it feasible? Is it appropriate? How will we know whether it is achieved? What does it say about how the framers of the goal think about education? Commentary afterwards indicated that the governors were in fact quite aware of what they were doing by including this goal: They wanted to have one goal that was direct, explicit, and a real challenge in order to express their collective agreement and determination. This was an instance of a rare focus on what was thought to be a very clear *outcome* that could be measured. But it was not the literal content of the goal, but the spirit that it communicated that was important to them—a decidedly political purpose. Nevertheless, the Third International Mathematics and Science Study (TIMSS) was initiated by the U.S. Department of Education and NSF in 1995 explicitly to measure the degree to which this goal had been achieved.

Another initiative formulated in this period is set forth in *Pathways to Excellence: A Federal Strategy for Science, Mathematics, Engineering, and Technology Education*, published in the waning days of the Bush administration in January 1992.[14] The *Pathways* document sets forth the strategic outline of an interagency effort involving NSF, Department of Education, National Institutes of Health, Department of Energy, Department of Defense, NASA, and other smaller agencies. However, the Clinton administration had a somewhat different set of priorities. While the *Pathways to Excellence* program proceeded, it lost some of its interagency features and became largely an NSF science education effort. The Clinton administration, with Congressional support, added two goals to the original six, downplayed the "first in the world" goal, and added a new thrust for technology in education that became identified with the leadership of Vice President Gore. The Clinton administration's program also shifted away from the *Pathways* focus on SMET education and extended activities to include education more broadly and work-force training programs in the Department of Labor.

A glimpse at *Pathways to Excellence* reveals another interesting phenomenon: the use of generic strategies in new contexts. NSF education programs on the K–12 and undergraduate levels have concentrated on two activities: teacher education (also called teacher enhancement for elementary and secondary education or faculty development for college and university teachers) and curriculum development. These activities were invented in the early days of NSF (e.g., the teacher institutes) and have continued via several iterations of program change over the last 30 or so years. It may seem surprising that there should be such constancy in approach while the environment and issues for science education have changed so much. However, teacher education and curriculum development continue to be seen as popular and effective strategies both by politicians and by professional education personnel. As the context changes, the rationale changes even as the activities remain more or less the same.

At NSF, the Directorate for Education and Human Resources (EHR) has adopted new designs for many of its programs even as their activities continue to focus on teacher education and curriculum development. Melvin George points out (see Chapter 1) the diagnosis that the problems of SMET education were systemic and called for "efforts that are broader than improving individual courses and individual laboratories in individual institutions." The keyword is *systemic* and the strategy is set forth, for its undergraduate SMET education programs, in the NSF report *Shaping the Future*.[15] The activities may seem to be the same, but the context and the rationale are new.

Subsequently, the Government Performance and Results Act (GPRA) was enacted early in the Clinton administration and has been superimposed on all federal programs and is part of the administration's "reinventing government" initiative. GPRA emphasizes outcomes. It calls for strategic plans by each agency and respective performance plans based on each agency's GPRA strategic plan. Thus, GPRA has been a significant driver of increased impact assessment and evaluation of NSF programs as well as those of other agencies.[16]

The NSF FY1999 GPRA Performance Plan lists five Outcome Goals, two of which pertain to education[17]:

- A diverse, globally oriented workforce of scientists and engineers
- Improved achievement in mathematics and science skills needed by all Americans

The first of these has been an outcome goal for the entire 50 years of NSFs existence. There is a robust delineation of this "work force" and there are substantial resources committed to monitoring it in both NSF and the Department of Labor. The other goal, usually referred to as the science literacy goal, needs a lot more definitional work before it is clear.[18,19] By and large, all NSF education programs include both, but it is frequently difficult to separate them in practice. To what extent is calculus reform one or the other? How do the dollar allocations break down to one and the other? This is a two-edged sword for evaluation. On the one hand, it creates a grand opportunity for evaluation to serve policy. On the other hand, it is a combination sure to create conflict between sponsors and evaluators who cannot agree on fundamental program philosophies.

The point of the foregoing recapitulation is twofold. First, federal support for education programs at NSF, the U.S. Department of Education, and other agencies is the product of a complex political process very hard to follow, even by the participants. The major shifts in priorities about education at NSF that occurred around the time of Sputnik had deep roots. The associated tensions recurred from time to time with greater or lesser intensity. Meanwhile, policy shifts at the top of government play themselves out within agencies and their staffs like the aftermath of a storm. Confusion ensues, exemplified through:

- Use of terms not well defined (e.g., science literacy)
- Use of policy research and scholarship—sometimes correctly, sometimes misleadingly—to rationalize program actions
- Professional judgment sometimes astute, sometimes not
- Trade-offs among interest groups
- A fondness for "experiments" without provision in program design to do the implied research

This situation diverts attention from the fact that there are insufficient resources to do the implied job fully. This is all done under tight program resources for staff and staff support, so that timing, coordination, and follow-through may be adversely affected and always put off until sometime later. Evaluation, when it occurs, is usually after programs are designed and implemented and it is too late to develop robust research designs for evaluation. Then, as time elapses and memories fade, there is the question of why the program funded yesterday does not fulfill the policy orientation of today.

My second point is what I call the *rhetoric of inflated expectations*, the most difficult challenge of all for evaluations. Implicit in the foregoing discussion is the sense of crisis, the aura of fear that accompanies policy rhetoric about education and its reform. For example,

- 1957: Sputnik stimulates policy to catch up with 10-foot-tall Russians at once.
- 1983: *A Nation At Risk* associates the dismal condition of the U.S. education system with the level of crisis associated with a "declaration of war."
- 1987: First in the world in science and mathematics by the year 2000, and the continual expectation of visible increase of student achievement scores on standardized tests that is hardly ever feasible (to detect or to occur at all) as a result of federally supported projects and programs.

This rhetoric, reinforced by letters like Senator Jake Garn's, leads (perhaps forces) agencies and their staff to make unrealistic claims for what their programs will do. The biggest problem for evaluation and evaluators comes from the rhetoric of inflated expectations coupled with the nature of NSF programs. NSF, as an agency of the federal government, takes great care to refrain from prescribing particular solutions. More properly, NSF sees itself as identifying broad problems and presenting opportunities.[20] The substantive work done is specified in the individual proposal for a project to be supported in the broad program. Thus, in many respects, the evaluation most important to NSF is the review it does at the front end that leads to the award. Subsequently, outcomes are defined at the project level. Indeed, most NSF resources are used at this *front end* with very little available for follow-up, staff site-visits, or oversight.

Nevertheless, when policymakers (e.g., members of Congress) want to know the outcomes, they frequently think in terms of, say, improved test scores measuring student learning, hinted at but never really promised in the program rhetoric. So at the agency level, the program promises to reform science and mathematics education. And this is passed along to projects as the agency program is inclined to fund those projects that promise what the funding agency wants to hear. As House[1] puts it, there is no way the programs and projects can live up to actual and implied promises, so we get the sad situation reported in *Science* magazine.[21] Quoting from this article: "But after 7 years and almost $600 million spent on the three [systemic] programs, officials are still a long way from knowing whether systemic reform works—or even what constitutes success."

9. THE CASE OF CALCULUS REFORM AND MATHEMATICAL ORTHODOXY

Calculus was invented independently by Newton and Leibniz in the latter third of the seventeenth century.* Isolated results of various aspects of the calculus accumulated over previous centuries, but Newton and Leibniz provided a new synthesis that gave calculus a power that was far greater than the individual features it synthesized. From the beginning, the calculus of Newton and Leibniz presented problems of *proof*, issues regarding its logical foundations, in relation to the orthodoxy of the day modeled by Euclid's *Elements*. Nevertheless, calculus was quickly accepted by scientists and mathematicians as a powerful tool that drove great progress in the science of mechanics for the next 300 years. The

---

* The use of the term *invented* here violates the accepted belief among most mathematicians that mathematical ideas lead lives of their own and therefore are discovered, not invented. This belief affects how mathematicians view (or accept) calculus reform, as we discuss later.

problems of proof and logical rigor were reconciled satisfactorily in the nineteenth century through the work of mathematicians like Cauchy and Weierstrass.

Meanwhile, in the nineteenth century, the classic curriculum of Greek, Latin, and mathematics dominated colleges in the United States. It was usually justified in terms of "mental discipline" and "faculty psychology." The mind was held to be composed of various faculties (e.g., memory, will, reasoning) that could be strengthened by mental exercise in somewhat the same way that muscles can be strengthened by physical exercise. Thus, mathematics was claimed to be crucial for promoting key mental faculties apart from its practical or explicit utility.[22] But by the close of the nineteenth century, advances in knowledge and the industrial revolution led to significant changes in our ideas of education, including the birth of the research university (notably Johns Hopkins University in 1876 and the University of Chicago in 1892), the rise of professionalism through education and scholarship in engineering and science, the rise of the public high school in the United States, and a focus on the study of subject matter of mathematics for its intrinsic value and not necessarily to strengthen mental faculties.

Calculus became the cornerstone of the new college curriculum in science as well as mathematics. What it was like and how it was taught is reflected in a textbook typical of the first half of the twentieth century, *Elements of the Differential and Integral Calculus* by Granville, Smith, and Longley (the second edition of which was published in 1911, with reprints continuing into the 1950s). This textbook presents calculus in "cookbook" style, with lots of drill and practice. In college, this one-year course in differential and integral calculus was usually preceded by more or less standardized courses with the titles College Algebra, Trigonometry, and Analytic Geometry. Then, following World War II, there was a major move to upgrade an "obsolete" mathematics curriculum.

Generally, the progression was toward greater formalism or, as mathematicians might put it, rigor through formal proof. The *function* concept became the cornerstone of the precalculus mathematics that replaced College Algebra and Trigonometry. Analytic geometry was combined with calculus, epsilons and deltas were introduced along with formal definitions of concepts like *limit*, and formal proof, especially of the mean value theorem, became de rigueur. A popular textbook was *Calculus and Analytic Geometry* by George Thomas, first published in the 1950s and still in print today. The audience and market was a more or less captive audience of students in the sciences and engineering, very few of whom were prospective mathematics majors. Calculus became a "rite of passage" in the sense that it was "the foundation for the study of the natural sciences, engineering, economics, and an ever-increasing number of the social sciences."[23]

Calculus is a substantial workload for every college and university department of mathematics in terms of course enrollments, although most enrollments in the calculus course are, as indicated, not mathematics majors. Therefore, mathematicians consider calculus a part of their "service load." In fact, many large universities with high calculus enrollments encourage prospective mathematics majors to take the "honors" version of the course. Demanding as calculus is for most students required to take it, those identifiable as prospective mathematics majors are a special group who were treated to even greater rigor. Many research universities have two or more versions of calculus, including regular and honors. This practice reflects a disproportionate distribution in the faculty resources assigned to the different versions of calculus, since the numbers at the honors end are very small compared with the gigantic enrollments in the "regular" calculus course.

During this process of evolution of the calculus course, the trend was toward greater rigor. Through most of history and into the nineteenth century, most mathematicians were practitioners of both pure and applied mathematics as was exemplified by, say, Gauss and his student, Riemann. Moreover, science and mathematics had interacted substantially since the time of Newton. But toward the end of the nineteenth century, mathematics became more clearly defined as an independent discipline and the proliferation of mathematical scholarship led to efforts by mathematicians to emphasize the unity and integrity of mathematics. This led to the perspective (belief system, in House's sense) that is given a nicely workable description by Holt.[24] Holt talks about the mathematical theorist Nicolas Bourbaki in whose work "rigor and abstraction reign." Holt also asserts that "some mathematicians have been hostile to Bourbaki's philosophy. By neglecting physical intuition and problem solving, they felt, it divorced mathematics from the real world, making it the subject a kind of logical theology." Rowe,[25] also identifying Bourbaki as the beginning, uses the term *Platonist* to identify a major belief system of mathematics. Rowe warns that "even if a mathematician professes not to be a Platonist, one should exercise considerable caution before accepting him at his word."

Platonism affects teaching in the sense that the mathematician as teacher believes he has the duty to carry forward with the aforementioned rigor and abstraction. In teaching, there is only one objective and that is to communicate the singularly true structure or nature of mathematics at hand. This carries over to an implicit conviction that there is a "right" way to teach a mathematical subject and for the student to learn it. This right way is not totally rigid, but it does have definite boundaries and minimal requisites, rationalizing the use of the term *belief system* in this context.

These developments of calculus and its teaching were occurring in the 1950s and 1960s and continued into the 1970s, when enrollments in higher education were increasing in both absolute and relative terms. Specifically, a larger proportion of the age cohort was going to college, growing from about 10% of the cohort in 1940 to about 50% by 1975. The great increase in the numbers and proportion of the age cohort attending college led to the perception of calculus as a "filter." Sheldon Gordon asserts (see Chapter 6) that "enrollment in mathematics drops by about 50% in each successive course from ninth grade algebra on up." This means that about half the students who enroll in calculus fail to complete the course with a grade C or better, many of these dropping out or failing altogether. While professors of mathematics in the 1970s and 1980s lamented this carnage, many saw it as inevitable. Their job was to teach *mathematics*, even if many students were not up to the demands.

This situation was the impetus for calculus reform: detachment from applications through an emphasis on formalism and mathematical rigor combined with a poor rate of success of students who enrolled in calculus. *Calculus: A Pump, not a Filter* was the rallying slogan. An important development that has promoted a new look was computer-based technology. While everyone was growing accustomed to the idea that computers crunched numbers, it suddenly occurred in the early 1980s that computers could also be programmed to crunch symbols, i.e., algebra. In fact, computers could find derivatives and integrals and, in many cases, carry out all the drill and practice one might find in Granville, Smith, and Longley or even in Thomas.

Calculus reform is often identified with a program initiated by NSF after the release of the Tulane Conference report.[26] At the time of the report, NSF was not providing much support for undergraduate education. The $15 million arising from the Garn letter (see Section 7 above) went largely to elementary and secondary teacher education programs. However, the Neal Report to the National Science Board (NSF 86-100) that called for renewed funding of undergraduate education in the sciences arrived at a political moment ripe for action, and calculus reform presented an interesting opportunity. Beginning in 1987 and continuing through 1995, NSF invested a total of approximately $43 million on the calculus reform effort. This effort was carried out through 354 individual project grants made during this period. About $22.4 million went to course and curriculum development (CCD) while $20.7 million went to non-CCD projects.[20]

10. EVALUATION OF CALCULUS REFORM

David Mumford[27] provides a superb and succinct rationale for calculus reform and clearly describes a central issue: that it is feasible to communicate mathematics *honestly* while "giving no proof." He sees room for memorization and drill as a pedagogical method in mathematics and for many examples, numerical and visual, based on things already familiar to the student. He speaks of a widespread disinterest in proof among scientists other than mathematicians as well as among nonscientist scholars and believes it is one reason why mathematics is "the most isolated of the sciences." Accordingly, he sees the aim of calculus reform as an opportunity to help students who are to be "scientists, engineers, and people in the world of affairs ... understand better what calculus is about or what it is good for." Moreover, he argues that this can be done honestly and without "the rigorous definition of limit with epsilon and delta and the mean value theorem." This is not the only approach or rationale among calculus reformers, but it is probably one that all share to a greater or lesser degree.

In addition, there are four national studies that address evaluation of calculus reform and several additional local studies. The four[28] include three published by the Mathematical Association of America and one to appear. The Tucker and Leitzel study, *Assessing Calculus Reform Efforts: A Report to the Community*, states in its executive summary:

- A key finding of the assessment study has been that *how* calculus is taught has changed more than *what* is taught.
- Changes include the use of graphing calculators and computers, open-ended projects, extensive writing, more applications, and the use of cooperative learning groups. Institutions report ongoing, lively discussions among faculty about approaches to topical material, modes of instruction, and ways of assessing student knowledge. Many faculty are concerned with getting the "right mix" of hands-on and technology-assisted work. The lecture method is being questioned and modified, and in some classrooms, abandoned.
- Faculty concerns have centered around the extra time needed to prepare for non-traditional instruction. After teaching reform calculus a few times, the preparation time is reduced but it is still more than in traditional courses.
- Information about the response of client disciplines [e.g., natural sciences, engineering, economics, and many of the social sciences] is spotty since reform is just beginning at most institutions, although feedback has been generally positive.

Tucker and Leitzel estimate that 150,000 students were enrolled in some form of reform calculus in 1994, which is 32% of the total college and university enrollments that year. There was estimated an additional 250,000 high school students enrolled that year in advanced placement calculus, of whom about half went on to take the Advanced Placement Calculus Exam—which now requires graphing calculators. Reform calculus is seen by these reports as still ongoing. Some assert that the reform will continue to spread, while others do not comment on its destiny.

However, calculus reform has not been embraced without dissent. Others have evaluated it and found it deficient. One of the four reports notes: "It must be acknowledged . . . that some college and university mathematicians believe that the increased use of technology, the introduction of more applications, and the increased emphasis on student communication is a change in the wrong direction." The evaluation of calculus reform needs to include this dissent and its actual and potential impact on the calculus reform movement now and for the future.

The Joint Mathematics Meeting in Orlando, Florida, in January 1996 included a special session on calculus that was to consider reports, like the Tucker and Leitzel *Report to the Community*, of reform calculus activity and status. The report of this meeting appeared as news in *Science*[29] under the heading "Calculus Reform Sparks a Backlash." The critics of reform argue that making calculus more appealing implies the risk of teaching to the lowest common denominator, "dumbing down" the subject, and making concessions that sacrifice orthodox standards of rigor. The critics of reform are particularly worried about computers and calculators. They see an overdependence on these instruments as working against the orthodoxy of calculus teaching and learning discussed previously (see Section 9).

In a direct response to Mumford, Klein and Rosen,[30] see calculus reform as a "facile response to a real problem: the declining performance of American students in college calculus courses." They scathingly reprove Mumford's argument and offer greater rigor and proof as the appropriate approach. They quote Abraham Lincoln to support their position (even though the quote is really a defense of the mental discipline and psychology of mental faculties that has lost some of its influence since Lincoln used it in the nineteenth century). In other words, Klein and Rosen provide an evaluation of calculus reform from the Platonist perspective and they definitely do not like what is happening in *calculus reform*.

The conflict within the mathematics community is part of a larger conflict referred to as the "science wars" being fought in academic professional journals from *Lingua Franca* to *Science* and the *Notices of the American Mathematical Society* as well as in business and literary journals like *The Economist and the New York Review of Books*. It also includes the public as is reflected, for example, in the California Mathematics Wars. Letters in the *Notices*[31] state that "mathematics education is working towards destroying the underpinnings of mathematics." Numbers of mathematics majors are dropping dramatically in many institutions, yet the mathematicians (or at least some of them) see this as a result of poor preparation of entering college students. This is a reflection of the Platonist belief system that only mathematicians can do mathematics education, only mathematicians can do history of mathematics. Thus, as Rowe suggests, a rigid separation between the scientific and humanistic cultures, the ideas of history, psychology, or education have no value for mathematics teaching and learning unless brought to bear by the mathematician. Therefore, the Tucker and Leitzel finding about calculus reform that "*how* calculus is taught has changed more than *what* is taught" explains the backlash from those who have no use for the "how" (humanistic) culture of calculus teaching and learning.

So what next? Is reform calculus a success? Is the number 150,000 of students enrolled in some form of calculus reform adequate in relation to the effort put into it? Is the 150,000 enrollment a peak? Will it grow? Some sort of insight into potential or actual answers to the questions requires further analysis along two lines. One is the issue of *belief system* and the backlash that are major barriers to further dissemination of calculus reform. Confronting this issue appears to be difficult, perhaps intractable. The Platonist critics of reform appear to be quite negative. They are against calculus reform and they seem to justify this with broad statements about what they do not like (e.g., dumbing down). At the same time, they do not seem to be coming up with concrete examples of what they find acceptable, say, an exemplar textbook to replace a reform text. In other words, there does not seem to be room for negotiation.

The other issue is the response of client disciplines to calculus reform. It is quite disappointing that so little information is available regarding the response of the client disciplines. This information is crucial to evaluation of calculus reform and it is worth the effort to get it. In view of the relatively high proportion of calculus enrollments that come from these client disciplines, the response from that quarter has potentially grave implications for the future of mathematics at the undergraduate level.

11. EVALUATION STRATEGIES

In light of the preceding discussion, it is important to outline some general observations about standard evaluation strategies.

11.1. The Gold Standard: Experimental Design

The gold standard for program evaluations is experimental design. This has been so for quite some time. A strong case for it has recently been reiterated by Robert Slavin[32]:

> ... the very core of social science [is] the use of control groups in program evaluations. It is odd to have to defend a procedure that is so well validated, widely used, and respected. Is it necessary to point out that the control group comparison is the quantitative design of choice, emphasized in every research methods text in every branch of science and social science? In every scientific field, the control group is the hallmark of rigor in experimental research. There are other rigorous alternatives in some circumstances, but no scientist would argue that an experimental–control comparison between equivalent groups is inappropriate or unscientific.

Of course, experimental design is what Slavin says it is, which is why it is referred to here as the gold standard. But a twofold problem often arises: (1) it is difficult—sometimes impossible—to implement and (2) it is costly, often extraordinarily so, in relation to the knowledge yield. A core issue is fulfilling the requirement of "comparison between equivalent groups"—how to define "equivalent" and identify groups that qualify. Of course, if an opportunity ever presents itself for a strong implementation of this method, then it probably is the strategy to use. But even then, if the underlying issue is not significant enough, it may not be worth doing it. The fact is that in the practical world of evaluation, experimental design often is not feasible.

This conclusion is based on the earlier discussion of how programs originate and are implemented. The programs are implemented first, typically without provision for evaluation—let alone experimental design. It is hard to even consider implementation of an experimental design unless those carrying out the treatment cooperate with the evaluator. Unfortunately, this often is constrained by concerns and protections about the burden on the project being evaluated. For example, informed consent and rules for protection of human subjects are not usually involved in an education project supported by an NSF grant. However, imposing an experimental design into the situation may well mean confronting these very reasonable requirements associated with research on human subjects. This changes the nature of the project substantially and the project directors may not want to go along. Moreover, it may not be feasible to do so—even if the project directors are willing and the requisite resources are present.

11.2. The Silver Standard: Quasiexperimental Design

Slavin does agree that there are rigorous alternatives in some circumstances. Some can be found in two valuable books: Cook and Campbell[33] and Campbell and Stanley.[34] One such technique is discontinuity of regression, used in one effort to design an evaluation of the NSF Graduate Fellowships Program in 1975 or 1976. Using standard inputs, undergraduates selected for graduate fellowships were among the very best and brightest in terms of such indicators as Graduate Record Examination scores, grades in college, and letters of reference. Subsequently, graduate fellowship holders typically completed doctoral studies sooner, published more, and advanced in research careers more rapidly than nonfellows.

Therefore, the evaluation issue was as follows: Given that the fellowship winners were so well-qualified, to what extent could their exemplary careers be attributed to the fellowship, when they would in fact have been expected to excel without it? For technical reasons, the effort was inconclusive; the regression had to be linear for a discontinuity to establish a difference and there was difficulty in establishing this necessary linearity. However, despite failure of the technique to be informative in this situation, the technique was originally chosen because it did work in analogous situations (e.g., dean's list winners). This underscores Slavin's qualification of "some circumstances" in his mention of alternatives to experimental design. There are undoubtedly other resources, but the two references mentioned here are provided as examples that there are indeed, as Slavin suggests, rigorous alternatives. In fact, the Internet points to books on resampling methods and data analysis, short courses with titles like "Messy Data," and Internet discussion groups on imputing missing data. These Web sites suggest continuing study and refinement of many of the techniques and issues that are developed in Cook and Campbell[33] and Campbell and Stanley.[34]

11.3. The Bronze Standard

We have to drop substantially, in terms of intrinsic standards of scientific rigor, to find other strategies. However, some can be quite valuable and useful, especially when the gold and silver are unattainable. Foremost in this category is what is usually referred to as professional judgment. A typical example is a set of site visits by one or more teams of appropriately qualified professionals. One might either assess an individual project or provide a synthesis across all projects to produce generalizations applicable to the overall program. The idea is not to issue grades to projects, but to evaluate the effectiveness of the program in reaching its goals. Often these days, these goals are determined by the GPRA requirements imposed on the agency, not the awardees. Interviews of project staff, faculty members, and administrators are major activities of a site visit. Frequently, the customers (students in revised courses, for example) are included among those interviewed. It is hard to be rigorous under these circumstances, but evaluators can be attentive to representatives and typicality if they cannot meet everyone involved with the project.

Examining records, data, and evaluations carried out by the project may also be possible as part of a site visit or separately, but that can be intrusive and not considered appropriate. Issues of confidentiality and privacy clearly are present in this situation. These may be thought of as constraints on the evaluation, but the evaluation certainly must move with care in accord with applicable standards of professional ethics.

The Committee on Science, Engineering, and Public Policy (COSEPUP) of the National Academy of Sciences, the National Academy of Engineering, and the Institute of Medicine recently reported its study *Evaluating Federal Research Programs: Research and the Government Performance and Review Act.*[35] The central finding of the report is that the annual assessment of results of basic and applied research is feasible (contrary to the view that the long-range nature of basic research precludes annual measurement). The report recommends "expert review" on quality, relevance, and international benchmarking of the research. It specifically notes that it does not support the view of some involved in the study that it should be possible to provide annual quantitative measures of both basic and applied research. "Quantitative measures" suggest assessment strategies of the kind that might be included in the gold or silver standard. The conclusions of COSEPUP regarding assessment of basic and applied research suggest that there might be value and utility to what is being suggested here for evaluation of SMET education programs at NSF.

The second level of activities involves collection of data (e.g., via questionnaire) from the participants in the treatment provided by the projects, for example, the teachers in a teacher development project or the students in a new course or a course using newly developed materials or teaching approaches. Interviews in addition to or instead of questionnaires may also be useful. These activities can be carried out on site or at a distance. Evaluators may also use data collected by the individual projects, such as routinely collected student evaluations of courses. Sometimes there are local or system rules preventing access to these data, but often they can be shared under appropriate circumstances. Of course, this level of activity can be combined with professional judgment, as previously discussed.

The third level consists of appropriately designed data collection from a secondorder audience, such as the students of teachers in a teacher development program. Policymakers frequently ask how we know whether teacher enhancement leads to improved student achievement. While the feasibility of detecting this kind of impact is questionable, it is still important that the issue be faced at every opportunity.

Of course, there are several variations and combinations of the aforementioned activities that might be developed as appropriate to different contexts. Patton[36] is a useful reference on design, data sets, data collection, instruments, and approaches—along with many others. At the same time, these remarks about "gold," "silver," and "bronze" are not intended to be a scientific primer on evaluations. The intent here is to provide a perspective, a conceptual orientation in the light of the realities of the political processes out of which NSF programs arise.

12. POLITICS AND EVALUATION

As previously mentioned, teacher education and curriculum development are generic activities that have been at the core of science education at NSF continuously for nearly 50 years. This is by no means a criticism, as these activities have evolved into many variations in response to changing needs and problems. Moreover, the contexts in which these activities have been supported have changed substantially over time from institutes to statewide systemic initiatives and from local course improvement to institutionwide reform. Nevertheless, there is a substantial political dimension behind these activities. The programs not only arise from policy (politics at the top of government), but also from the politics of program design and implementation. Moreover, professional constituencies like these activities and lobby for them. Agency turf issues all play themselves out in this process. As evaluations are also a part of this milieu, evaluation research designs, relationships between evaluators and agency sponsors and their constituencies, and the perceived value and merit of the evaluations themselves are also affected. Again, the intent here is not to pass judgment, but to point out the nature and sources of problems for evaluators and evaluations.

To sharpen this point, consider an example from the past. In the early 1970s, the extraordinary growth of the science and technology enterprise, of higher education, and of the population in general began to slow and level off. Some thought this was a temporary condition associated with the Vietnam War while others thought that it was deeper and more structural, i.e., another instance of exponential-like growth processes coming to an inevitable end. There was the Club of Rome Report, for example, and concern about unemployed aerospace engineers as the space program was reduced following the successful Apollo program. There was also a great deal of attention given in the scholarly and popular media about overeducated U.S. Ph.D.s pumping gas and driving cabs, and whether a college education really provided a return commensurate with the investment. Elliott Richardson commissioned studies as Secretary of Health, Education, and Welfare. The famous Newman Report[37] chided higher education for its sameness. If you look at catalogs from two- and four-year colleges, and from universities, he said, there is a similarity in purposes, courses, and programs. Such differences as might exist were not visible on the surface as presented in the catalogs. The report became known for its recommendation of "less time, more options"—fewer requirements and more variation in higher education programs. This, in fact, had an enormous policy impact. The federal government essentially cut out support to institutions and faculty for program development and replaced it with direct student support in the form of loan subsidies and tuition grants.

Perhaps the persistence of generic activities like teacher education and curriculum development inhibit out-of-the-box thinking and the examination of the full range of options that might exist to address problems and issues in education. Instead of dressing up (and dragging down) teacher education and curriculum development by embedding them in vague and complicated management systems like systemic initiatives, policymakers, policy wonks, and their existing constituencies might provide stimuli that extend the range of program options they consider.

This is not an original thought. A military training model has been described[38] as a possible response to government concern about science education and the work force. This proposal speaks of "tuition waivers, monthly stipends, and paid summer work experience" (p. 120) for young people to become educated in SMET fields. While hypothetical, it is at least an example of out-of-the-box thinking (like the Newman Report in its day) that should be part of the policy discussions stimulated by evaluation results.

As for calculus reform, Rowe[25] argues that we are in a "trend away from a monolithic picture of mathematics [such as is reflected by Bourbakism] to a more pluralistic outlook reflecting a broad range of diverse mathematical interests." Analogous to Rowe's challenge facing the history of mathematics, perhaps mathematics education, including the teaching of calculus, can benefit from "a constructive dialogue between the parties representing these differing interests" and look for grounds for peaceful coexistence among pluralistic views rather than warfare over a monolithic belief system.

13. THE WISDOM OF POOH

In this difficult program initiation and evaluation milieu, a cautionary tale I call Pooh's Assessment Dilemma is enlightening. In one of the Winnie-the-Pooh stories,[39] Christopher Robin, Pooh Bear, and Piglet were talking when Christopher mentions casually that he had seen a "Heffalump." "What's a Heffalump?" Pooh and Piglet were wondering to themselves when Piglet says that he's seen one, too. Pooh comments, "I've seen one, too, I think." On that vague and uncertain basis, Pooh tells Piglet a little later that he has decided to catch a Heffalump. After Piglet had insinuated himself into the search, as Pooh had wished he would without being asked directly, the discussion turned to how to catch a Heffalump. It was decided to dig a Very Deep Pit for the Heffalump to fall in. There ensued a discussion about when and how, through which Piglet and Pooh were mutually reassuring on various sticky points. They talked about where the trap should be dug, how deep it should be, and how to dig the trap close enough to where the Heffalump was, but not so close that it would see them digging it. Finally, the issue of bait came up. Each thought of bait that the other would have to provide before Piglet suggested that he would dig the pit if Pooh would get honey for bait.

So with a rather arbitrary and capricious resolution of the details, the Very Deep Pit is dug, Pooh's last very large jar of Hunny is set in it about halfway down, and Pooh and Piglet go home to sleep. But Pooh worries and worries about the situation—the Very Deep Pit, the Hunny, its placement. So he goes down into the Very Deep Pit several times to taste the Hunny. Finally, on one trip, he gets his head stuck in the very big Hunny jar and falls into the pit.

In the morning, Piglet rushes over to the Very Big Pit, sees this creature in it with a Very Big Head and decides it is a Heffalump. Whereupon, he dashes off in fear to find help with the Terrible Heffalump in the Very Deep Pit. When Christopher Robin gets to the Very Deep Pit, he recognizes Pooh and helps him get his head out of the Hunny jar. Piglet is mortified by his foolishness, but all are comforted by going off to breakfast together.

It may take the fun out of playing in the woods, but the tale has a few obvious pointers for evaluators:

1. Be clear about what you are looking for and have a common understanding about it with the evaluation sponsor.
2. Use appropriate traps to "capture" the necessary information.
3. Go beyond hunches and personal preferences in selecting "bait."
4. Be cautious about identifying what you think you have found.

Most important, don't get so carried away by engaging in the activity that you lose sight of the fact that a product and conclusion are expected. Finally, be prepared to explain to the sponsor, if you fail to find what he thinks should be there, whether it really isn't there or whether your trap just wasn't up to catching it.

REFERENCES

1. E. R. House, "The Politics of Evaluation in Higher Education," *Journal of Higher Education* XLV(8)(1974):618–627.
2. V. Bush, *Science—The Endless Frontier* (National Science Foundation, Washington, DC, 1945). [Reprinted by the National Science Foundation in 1960, 1970, 1980, and 1990 to mark the decennial anniversaries of its establishment.]
3. D. J. Kevles, "The National Science Foundation and the Debate over Postwar Research Policy, 1942–1945: A Political Interpretation of *Science—The Endless Frontier*," *Isis* 68(March 1977):5–26.
4. J. M. England, *A Patron of Pure Science: The National Science Foundation's Formative Years, 1945–57* (National Science Foundation, Washington, DC, 1982).
5. H. Aaron, "Budget—The Big Picture" [Editorial], *Science* 281(1998):345.
6. H. M. Sapolsky, "Financing Science after the Cold War," in D. Guston and K. Kennison (eds.), *The Fragile Contract: University Science and the Federal Government* (MIT Press, Cambridge, MA, 1994).
7. H. Kreighbaum and H. Rawson, *An Investment in Knowledge: The First Dozen Years of the National Foundation's Summer Institutes Programs to Improve Secondary School Science and Mathematics Teaching 1954–1965* (New York University Press, New York, 1969).
8. T. W. Schultz, "Investment in Human Capital," *American Economic Review*, 1961 [reprint of the 1960 Presidential Address to the American Society of Economists].
9. G. Becker, *Human Capital: A Theoretical and Empirical Analysis*, 3rd ed. (University of Chicago Press, Chicago, 1993) [1st ed. 1964, 2nd ed. 1975].
10. National Center for Education Statistics, *Digest of Education Statistics 1996* (U.S. Department of Education, Washington, DC, 1996) Table 155, p. 151.
11. M. Lomax, *A Minor Miracle: An Informal History of the National Science Foundation* (National Science Foundation, Washington, DC, 1976).
12. National Commission on Excellence in Education, *A Nation at Risk, The Imperative for Educational Reform* (U.S. Department of Education, Washington, DC, 1983).
13. General Accounting Office, *New Directions for Federal Programs to Aid Mathematics and Science Teaching*, GAO/PEMD-84-5 (U.S. General Accounting Office, Washington, DC, 1984).
14. Committee on Education and Human Resources of the Federal Coordinating Council for Science Engineering and Technology, *Pathways to Excellence: A Federal Strategy for Science, Mathematics, Engineering, and Technology Education* (Office of Science and Technology Policy, Executive Office of the President, Washington, DC, 1992).

15. National Science Foundation, *Shaping the Future: New Expectations for Undergraduate Education in Science, Mathematics, Engineering, and Technology. Report of the Advisory Committee for Undergraduate Education* (National Science Foundation, Washington, DC, 1996).

16. J. A. Hoskinen, *Statement on Government Performance and Review Act of 1993*, Testimony for the Office of Management and Budget before the House Committee on Reform and Oversight, February 12, 1997.

17. National Science Foundation, *National Science Foundation FY 1999 GPRA Performance Plan*, 1998.

18. F. J. Rutherford and A. Ahlgren, *Science for All Americans* (Oxford University Press, London, 1990).

19. M. Shamos, *The Myth of Scientific Literacy* (Yale University Press, New Haven, 1995).

20. W. E. Haver, *Calculus: Catalyzing a National Community for Reform, NSF Awards 1987–1995* (Mathematical Association of America, Washington, DC, 1998).

21. J. Mervis, "Mixed Grades for NSF's Bold Reform of Statewide Education," *Science* 282(1998):1800–1805.

22. D. L. Roberts, *E.H. Moore's Failed Attempt to Involve the American Mathematical Society in Pedagogy*, Presentation at the American Mathematical Society Meeting in Philadelphia, April 1998.

23. A. C. Tucker and J. R. C. Leitzel, *Assessing Calculus Reform Efforts: A Report to the Community* (Mathematical Association of America, Washington, DC, 1995).

24. T. Holt, "The Joy of Sets: Who's Responsible for the New Math?" *Lingua Franca* (October 1998):76.

25. D. Rowe, "New Trends and Old Images in the History of Mathematics," in R. Calinger, Ed. *Vita Mathematica* (Mathematical Association of America, Washington, DC, 1996).

26. R. G. Douglas (ed.), *Toward A Lean and Lively Calculus* (Mathematical Association of America, Washington, DC, 1986).

27. D. Mumford "Calculus Reform for the Millions," *Notices of the American Mathematical Society* 44(5)(1997):559–563.

28. The four are: (a) Tucker and Leitzel, Ref. 23; (b) A. W. Roberts, *Calculus: The Dynamics of Change* (Mathematical Association of America, Washington, DC, 1996); (c) Haver, Ref. 20; and (d) S. L. Ganter, *Ten Years of Calculus Reform: A Report on Evaluation Efforts and National Impact* (Mathematical Association of America, Washington, DC, in press).

29. B. Cipra, "Calculus Reform Sparks a Backlash," *Science* 271(1996):901–902.

30. D. Klein and J. Rosen, "Calculus Reform for the $Millions," *Notices of the American Mathematical Society* 44(10)(1997):1324–1325.

31. J. Rosen, "Mathematics Education and Policy" [Letter], *Notices of the American Mathematical Society* 43(5)(1996):534–535.

32. R. E. Slavin, "A Rejoinder: Yes, Control Groups Are Essential in Program Evaluation: A Response to Pogrow," *Educational Researcher* 28(3)(1999):3638.

33. T. D. Cook and D. T. Campbell, *Quasi-experimentation: Design and Analysis Issues for Field Settings* (Rand McNally, Chicago, 1979).

34. D. T. Campbell and J. Stanley, *Experimental and Quasi-experimental Designs for Research* (Rand McNally, Chicago, 1966).

35. COSEPUP, *Evaluating Federal Research Programs: Research and the Government Performance and Results Act.* A report of the Committee on Science, Engineering, and Public Policy; National Academy of Sciences, National Academy of Engineering, and Institute of Medicine (National Academy press, Washington, DC, 1999).

36. M. Q. Patton, *Qualitative Evaluation and Research Methods*, 2nd ed. (Sage Publications, Newbury Park, CA, 1990).

37. F. Newman, *The Second Newman Report: National policy and higher education; report of a special task force to the Secretary of Health, Education, and Welfare*, Frank Newman, chairman (MIT Press, Cambridge, MA, 1973). [Note: The first Newman report was published in 1971 under the title: *Report on higher education*, by the U.S. Department of Health, Education, and Welfare.]
38. S. Tobias, D. Chubin, and K. Aylesworth, *Rethinking Science as a Career* (Research Corporation, Tucson, 1995).
39. A. A. Milne, *Winnie-the-Pooh* (Dutton, New York, 1926) (reprinted by Puffin Books of Penguin Press, 1992).

CHAPTER 10

Crossing the Discipline Boundaries to Improve Undergraduate Mathematics Education*

James H. Lightbourne, III

The purpose of this chapter is to indicate how changes occurring in undergraduate science and engineering education can inform and support improvement in undergraduate mathematics education. Reports and discussions on education in sessions at national and professional society meetings have common themes and findings across the various disciplines. However, there is not much exchange of information across these discipline boundaries. Similarly, visiting college campuses, one frequently finds mathematics faculty who have more in common in terms of their views and practices to improve undergraduate education with faculty in physics, for example, than colleagues in the mathematics department. This lack of communication and collaboration across discipline boundaries results in missed opportunities that would benefit mathematics departments and mathematics education.

Section 1 provides a brief summary of the Tulane Conference recommendations and general trends found among the various calculus reform projects.

---

* The opinions expressed in this chapter are those of the author and are not intended to represent the policies or position of the National Science Foundation.

Section 2 provides similar information from reports on undergraduate education in chemistry, earth sciences, engineering, and physics. Section 3 describes projects in undergraduate science and engineering education that illustrate specific efforts occurring in these disciplines. Section 4 provides summary observations.

Material for this paper is drawn liberally from national reports and testimony to the National Science Foundation obtained during the *Shaping the Future* hearings.[1,2]

1. CALCULUS REFORM

The "Tulane Conference,"[3] with funding from the Sloan Foundation, was held in January 1986 in conjunction with the Annual Joint Mathematics Meetings. The conference, attended by 25 invited participants, identified five general problems encountered at that time in the teaching of calculus:

- Too few students successfully completing calculus
- Students performing symbolic manipulations with little understanding or ability to use calculus in subsequent courses
- Faculty feeling frustrated with poorly prepared, poorly motivated students
- Calculus being required as filter through which other disciplines culled out students but made little use of calculus in their courses
- Mathematics lagging behind other disciplines in use of technology

The mathematics community responded by developing new texts and other materials for teaching calculus. *Assessing Calculus Reform Efforts*[4] is a report on the findings of a Mathematical Association of America study to assess the calculus reform movement. Many of these calculus reform efforts have been supported by the National Science Foundation (NSF) through the NSF Calculus Program and other programs at NSF.[5] The materials offer a variety of approaches to teaching and learning calculus, reaching a broader student audience. Topics are presented through several representations; for example, graphical, numerical, symbolic, and written or verbal description. The changes in instructional practice include introduction—or increased use—of technology, modeling and applications, collaborative learning, student projects, student writing, and oral presentations.

2. REPORTS FROM OTHER DISCIPLINES

During this same period of time, other disciplines have been reconsidering how their subjects are taught. As the following examples illustrate, the concerns and recommendations for improvement parallel those in the mathematics community and, in particular, the calculus reform movement.

2.1. Chemistry

The report *Innovation and Change in the Chemistry Curriculum*[6] is the proceedings from an NSF-sponsored workshop held in May 1992. Attending the workshop were approximately 50 faculty representing two- and four-year colleges and universities. The workshop participants identified several areas of concern:

- Goals of instruction too highly focused on student mastery of specific content
- Students do not gain appreciation of scientific method nor have experience of discovery
- Student assessment designed primarily to be relatively inexpensive and to require minimum faculty and student time
- Inadequate use of modern instrumentation to engage students in science and discovery
- Student laboratories merely "repeat the instructions" experiments

Workshop recommendations included:

- Critically examine course goals and align assessment with these goals
- Give all students, whether they become scientists or not, a sense of professionalism and involvement, an appreciation of the scientific method and how it impacts on public discourse, and an understanding of research and the excitement of exploration and discovery
- Develop methods for assessing student learning that do not focus on just the lowest levels of learning
- Introduce modern instrumentation and computers that engage students in the process of discovery
- Educate all students in use of the scientific method and the importance of chemistry to the economy as well as advances in medicine and improvements in the environment
- Bring faculty research expertise into the design of laboratory courses illustrating the role of experimentation, observation, and scientific deductions in the laboratory while avoiding whenever possible the classical "repeat the instructions" experiments

2.2. Earth Systems Science

The American Geophysical Union in cooperation with the Keck Geology Consortium and with support from NSF convened a workshop in November 1996 to define common educational goals among all disciplines in the earth sciences. The workshop had over 50 participants in the earth sciences, representing two- and four-year colleges, universities, national laboratories, and professional societies. The findings of the workshop are recorded in the report *Shaping the Future of Undergraduate Earth Science Education.*[7]

The report was critical of current earth science courses, noting the following concerns:

- Prevalence of passive rather than active learning
- Emphasis on factual knowledge without experience in the process of science
- Cookbook rather than inquiry-based labs
- Lack of relevance of course materials

The recommendations of the report are to:

- Reaffirm the importance of classroom, laboratory, and field activities that encourage active inquiry
- Illuminate societal issues and the connections between scientific and nonscientific disciplines
- Decrease emphasis on fact-focused, lecture style courses
- Emphasize in-depth understanding of a few fundamental elements in the disciplines and the development of critical thinking
- Incorporate advances in learning theory to promote genuine inquiry, critical thinking, proficiency in written and oral communication, and lifelong learning skills into courses at all levels
- Use field studies to connect earth systems science content to student life experiences and involve students in questioning, data gathering, and interpretation
- Utilize technology to create virtual field trips, access to large spatial and time series databases, and exchange observations and ideas over large distances
- Inform students of value of research, create a research group environment, encourage student presentations of results to departments, institutions, regional or national meetings, encourage students to publish

A definition of student research. The report gave considerable attention to faculty and student research. Student research is defined as inquiry based study, rediscovery, or discovery resulting in original contribution. Student research includes hypothesis formulation, collection and use of real-time data and other research materials to test hypotheses, and analysis of data in individual or group settings. This definition of student research is intended to include a broad spectrum of discovery activities as appropriate for different content and student levels.

2.3. Engineering

NSF sponsored a workshop in June 1994, engaging 65 participants who represented engineering faculty, professional societies, industry, and students. The proceedings of that conference comprise the report *Restructuring Engineering Education: A Focus on Change.*[8] Concerns raised at the workshop included:

- Classes typically taught in large lecture settings
- Problem assignments and assessments are highly structured
- Lack of research base in teaching and learning
- Lack of attention to different student career goals

Proposed is creation of learning environments that include:

- Active, collaborative learning
- Use of modules
- Research, development, and practical experience for undergraduates
- Learning-by-doing, the norm in professional fields
- Increased integration of science, mathematics, and engineering sub-discipline content
- Recognition of different backgrounds and career goals of students
- Rigorous educational research base in teaching and learning
- Appreciation for the complexities of physical devices and structures

In terms of content, the workshop concluded that it is impossible to define an engineering curriculum applicable at all institutions. Each school needs to consider its own constituents and diversity of programs should be encouraged.

This flexibility in developing engineering programs is also reflected in the revised criteria used to accredit engineering programs by the Accreditation Board for Engineering and Technology, Inc. (ABET). Engineering Criteria 2000[9] was approved by the ABET Board of Directors for a two-year comment period that began in January 1996. A phased-in implementation began with the 1998/1999 visit cycle, with full implementation of the Criteria 2000 in fall 2001. The programs are evaluated based on student outcomes, with specific course requirements not stated to the extent done so in past years. Evaluation evidence that may be used includes, for example, student portfolios, including design projects; nationally normed subject content examinations; alumni surveys that document professional accomplishments and career development activities; employer surveys; and placement data of graduates.

In the new criteria, engineering programs are expected to demonstrate that their graduates have the following capabilities:

- Ability to apply knowledge of mathematics, science, and engineering
- Ability to design and conduct experiments, as well as to analyze and interpret data
- Ability to design a system, component, or process to meet desired needs
- Ability to function on multidisciplinary teams
- Ability to identify, formulate, and solve engineering problems
- Understanding of professional and ethical responsibility
- Ability to communicate effectively
- Broad education necessary to understand the impact of engineering solutions in a global/societal context
- Recognition of the need for and an ability to engage in lifelong learning
- Knowledge of contemporary issues
- Ability to use the techniques, skills, and modern engineering tools necessary for engineering practice

The changes in the ABET evaluation criteria—and the consequent changes in engineering education—potentially could have a significant impact on mathematics departments. As indicated earlier, the new criteria are based on expected student outcomes, rather than on a checklist of courses. This presents an opportunity for mathematics departments to work with the engineering departments toward the outcomes-based criteria. However, the new criteria also present the danger of losing the teaching of courses for those departments that do not do so.

2.4. Physics

There is a growing group of individuals conducting discipline-based research in teaching and learning in physics. The paper "Resource Letter on Physics Education Research"[10] provides an annotated compilation of over 200 references, primarily focused on postsecondary education. Many of the references are empirical studies that consider student understanding of a specific topic. "Teaching Physics: Figuring Out What Works"[11] is an example of a more general paper. This paper poses three questions: What is involved in understanding physics? What do students bring to physics classes? How do students respond to what they are taught? Among other findings, the paper reports results from a study comparing three educational environments: traditional (large lectures with small group recitations and laboratories), tutorials (including student group work on research based worksheets), and workshop physics (lectures, recitations, and laboratories combined into lab-based sessions).

A large body of research involves use of a multiple-choice diagnostic test, the Force Concept Inventory (FCI).[12] FCI is a 29-question test that assesses student understanding of concepts in mechanics. Studies (e.g., Redish and

Steinberg[11]) reported in the literature show that student performance on FCI does not necessarily improve with traditionally taught classes; in fact, student performance actually appears to have deteriorated. In addition, students appeared to deteriorate with traditional instruction in general areas such as learning independently, linking physics with reality and mathematics, and understanding concepts. An extensive study[13] conducted in a variety of school settings concluded that students in interactive classes consistently scored better on diagnostic tests than students in traditionally taught classes.

Testimony given during the hearings for *Shaping the Future*[2] reported that "we know beyond any reasonable doubt" that engaging undergraduate students in active learning and active research, in close contact with faculty and other students, encourages students of all kinds to continue toward a career in science. Students are engaged in active learning through several means:

- Classroom instruction that keeps students active
- Early participation in research
- Appropriate use of technology—for example, interconnected computers provide focus for small group discussions; spreadsheets provide means for numerical computation; digital video processing provides means to study realistic applications

Concern was expressed that (1) computer simulations of experiments easily conducted in the laboratory and (2) computer-aided instruction as was traditionally implemented isolate students and do not have desired outcomes. Also, the demand for coverage of material too often outweighs the demand for conceptual understanding and true learning.

3. EXAMPLES FROM OTHER DISCIPLINES

The following is a sample of projects that have recently been funded by NSF in disciplines other than the mathematical sciences. These illustrate in more specific terms some of the changes occurring in other disciplines.

- The chemistry department of San Jose State University in California is restructuring the quantitative analysis course to include topics in contemporary analytical problems and technology. Topics include multi-element and trace analysis, and combined physical and chemical characterization. These topics are explored in a laboratory designed to emulate the working environment of a modern commercial laboratory. The laboratory includes a computer-controlled HPLC, atomic absorption, voltammeter instrumentation, and photomicrography for physical characterization of solids.

- Grand Valley State University is developing a series of hands-on experiences that link geology and chemistry through the use of cathodoluminescent (CL) imaging. CL imaging provides a visual picture of trace element distributions in common minerals, which allows sedimentation and stratigraphy labs to observe the complex evolution of the rock record by recording the changes in the composition of minerals. Students also gain an understanding of the chemical processes that produce the complex compositional changes in minerals.

- Daytona Beach Community College is creating a new instructional environment for introductory courses in electronics, computer-aided design, civil engineering, and computer programming. The objective of the project is to develop a virtual classroom environment through which students can access course materials and interact with other students and faculty.

- The University of Florida is developing a real-time interactive flight test program. The program allows participants to perform airborne experiments, with data reduction and analysis in real time on board the aircraft. Videos and other sensor data from the aircraft are sent to classrooms via videoconferencing, so that the classes may actively participate in a real-time flight test.

- The conception, production, evaluation, and dissemination of a series of interactive modules for the teaching and learning of fluid mechanics in science and engineering is being developed jointly by Stanford University, MIT, and the University of Illinois. The modules focus on fundamentals of design but could be used in curricula of other disciplines. The primary objectives are to enhance student problem solving, intuition about complex flow phenomena, and retention of knowledge.

- Faculty in chemistry at the University of Massachusetts at Amherst are developing, for large enrollment classes, a Web-based electronic homework and intelligent tutoring system that is interactive and incorporates elaborate feedback, multimedia elements, and automatic recordkeeping. A tool set will allow the system to be used by both other universities and other disciplines, with testing planned at several cooperating institutions and departments.

- A project at Wright State University (Ohio) is addressing the underrepresentation of students with disabilities in the sciences by developing resources that can be used nationally to assist biological educators in colleges and in grades 7–12 in designing appropriate laboratory exercises for students with disabilities. A Source Book will assists educators in creating an effective natural/physical science laboratory for students with disabilities, and adaptive laboratory manuals are being developed in the biological sciences.

- A project at Indiana University at Bloomington provides field equipment for state-of-the-art training of field-based environmental geoscientists. The setting for this field instruction is a southwestern Montana watershed with headwaters from a large snow catchment and lower drainage in an arid structural basin. The project provides a suite of permanently deployed instruments, which are installed and calibrated by students with additional measurements taken along grid surveys in watershed subareas. Students work with portable laptop computers, GPS receivers, and a GIS system in data retrieval, analysis, and information display.

- Introductory Biology, Comparative Physiology, and the Human Physiology courses developed at Bowdoin College involve aspects of the experimental design and execution, including hypothesis generation, experimental design, data collection and analysis using modern equipment, and presentation of results in a variety of formats. By enhancing courses at all levels of the curriculum, the department is promoting an understanding of the process of science and the fundamental concepts of each level.

- A microcomputer-based laboratory at the University of Maine, Farmington is being used to introduce an inquiry-based curriculum into the general physics sequence. The student population of the course is approximately half science majors and half secondary education majors. In this project there is a particular focus on the preservice teachers in the course. This focus consists of having "alumni" of the general physics sequence return as peer instructors in both the workshop physics course and the conceptual physics course designed for nonmajors. This is being done by having the physics and secondary education faculty work together to affect program changes that would require the secondary education students to have this teaching experience as part of the degree requirements in education. The goals are to further improve the understanding of physics of these science teachers-to-be and to give them some practical experience with an inquiry-based physics curriculum.

- The Union College (New York) Environmental Studies program, which is offered through the Geology, Civil Engineering, Biology, and Chemistry Departments, is advocating a major new research and teaching directive entitled the Ballston Lake Initiative. The central focus of the initiative is Ballston Lake as an environmental system. This broad-based initiative includes a variety of class, field, and research projects that are interdisciplinary and multidisciplinary.

4. SUMMARY OBSERVATIONS

The previous sections serve to illustrate common concerns and recommendations for undergraduate education throughout science, mathematics, and engineering disciplines. In summary, the following major areas are identified:

- *Course Emphasis.* Current courses tend to emphasize manipulative skills, routine experiments, or cookbook techniques rather than student understanding and competence in the subject. Course design, including student testing, should place more emphasis on understanding of concepts, scientific method, and relevance in a broader context. The curriculum in general should reach a wider spectrum of students in terms of backgrounds, interests, and career goals.
- *Educational Practices.* In general, there is too much reliance on lecture, routine student exercises and laboratories, and exam designed primarily to minimize student and faculty time. The reports recommend that faculty increase their use of collaborative learning, discovery-based student activities and student research, projects, writing assignments, oral presentations, and other practices that provide more engaging and effective education.
- *Computer Technology.* Computer technology can be used, for example, to engage students in discovery, provide access to large databases, gather information, and collaborate over large distances.
- *Content.* In general, these reports conclude that courses try to cover too much material, at the expense of a sufficiently deep treatment of the subject. It is recommended that courses include fewer topics for which deeper student understanding would be possible and expected.
- *Research Based Educational Decisions.* Discipline based research in teaching and learning should provide a scholarly basis for informed educational decisions. This research is providing insights into what students actually learn and what educational practices are effective for improved learning.

At disciplinary society meetings across the country, faculty discuss ways to improve undergraduate education. Although these meetings are held within the separate disciplines, the issues, concerns, and recommendations have much common ground across the disciplines. These faculty return to their home institutions and too often find themselves working in isolation. Colleagues in their own discipline may not be receptive, and they do not communicate across the discipline and department boundaries.

The benefits of these interactions across discipline lines are multiple. Students benefit in having both content and pedagogical approaches, including uses of technology, reinforced in different courses and discipline settings. The content in mathematics courses can be enriched through applications relevant to the other courses that students take. Faculty implementing similar strategies in different disciplines can benefit through collaboration. The institutional support possible through having a critical mass of faculty with common interests is also not realized. Moreover, the mathematics department, in general, can be better positioned as central to the institutional mission to provide undergraduate education.

There are exceptions, campuses on which undergraduate education benefits from close working relationships across discipline boundaries. These situations do show that such interactions are achievable and provide models for such work. However, there needs to be many more of these exceptions; in fact, these interactions across the disciplines should become the norm.

REFERENCES

1. *Shaping the Future: New Expectations for Undergraduate Education in Science, Mathematics, Engineering, and Technology* (National Science Foundation, Washington, DC, NSF 96-139, 1996).
2. *Shaping the Future, Volume II: Perspectives for Undergraduate Education in Science, Mathematics, Engineering, and Technology* (National Science Foundation, Washington, DC, NSF 98-128, 1998).
3. R. G. Douglas (ed.), *Toward a Lean and Lively Calculus*, MAA Notes No. 6 (Mathematical Association of America, Washington, DC, 1986).
4. A. C. Tucker and J. R. C. Leitzel, *Assessing Calculus Reform Efforts* (Mathematical Association of America, Washington, DC, 1995).
5. W. E. Haver (ed.), *Calculus: Catalyzing a National Community for Reform* (Mathematical Association of America, Washington, DC, 1998).
6. *Innovation and Change in the Chemistry Curriculum* (National Science Foundation, Washington, DC, NSF 94-19, 1994).
7. M. F. W. Ireton, C. A. Manduca, and D. W. Mogk, *Shaping the Future of Undergraduate Earth Science Education* (American Geophysical Union, Washington, DC, 1997).
8. *Restructuring Engineering Education: A Focus on Change* (National Science Foundation, Washington, DC, NSF 95-65, 1995).
9. *Engineering Criteria 2000*, Engineering Accreditation Commission of the Accreditation Board for Engineering and Technology, 1995.
10. L. C. McDermott and F. Redish, "Resource Letter on Physics Education Research," preprint.
11. E. F. Redish and R. N. Steinberg, "Teaching Physics: Figuring Out What Works," *Physics Today* 52(1999):24–30.
12. D. Hestenes, M. Wells, and G. Swackhammer, "Force Concept Inventory," *Physics Teacher* 30(1992):141–158.
13. R. R. Hake, Interactive-engagement versus traditional methods: A six-thousand student survey of mechanics test data for introductory physics courses, *American Journal of Physics* 66(1998):64–74.
14. *Science Teaching Reconsidered: A Handbook* (National Academy of Sciences, Washington, DC, 1997).

About the Authors

Jack Bookman has been a member of the Duke University mathematics faculty since 1982. He has an M.A.T. degree from SUNY-Binghamton and taught high school mathematics for four years. From 1979–1981, he was a graduate student in mathematics at Duke University, earning an M.A. before joining the faculty as an instructor. Beginning in 1984, while teaching at Duke, he attended UNC-Chapel Hill, earning a Ph.D. in Education in 1991. From 1989–1994, he served as project evaluator for Duke University's *Project CALC: Calculus As a Laboratory Course*. He has also served as an evaluation consultant for calculus reform projects at Dickinson College and University of Puerto Rico. His current evaluation project involves examining how students learn in web-based environments using the materials of the NSF- funded Connected Curriculum Project, which is developing a World Wide Web library, of modular, interactive laboratory materials for lower-division mathematics (housed at Duke University). He is also involved in preparing future secondary and college mathematics teachers.

Alphonse Buccino is currently an independent consultant doing business as Contemporary Communications, Inc. in the areas of mathematics, science policy programs, science studies, mathematics and science education, program evaluation, technology applications, and e-commerce. Present or recent affiliations include Young & Rubicam, SRI International, the Brookings Institution, Johns Hopkins University, and George Mason University. He was dean of the University of Georgia's College of Education for over 10 years beginning in April 1984. Before going to Georgia, Buccino spent 14 years at the National Science Foundation serving as an expert and frequent spokesperson on the education and training of scientific and technical personnel. Dr. Buccino spent the year from March 1992 to March 1993 on a special assignment in the White House Office of Science and Technology Policy as advisor on mathematics and science education to the President's cabinet-level science advisor. Buccino holds a Ph.D. in mathematics from the University of Chicago.

Paul Davis has spent his career teaching and practicing applied mathematics at Worcester Polytechnic Institute. He is currently Dean of the Interdisciplinary and Global Studies Division and Director of WPI's London Project Center. He has been active in SIAM, most recently as a member of the Steering Committee for its Mathematics in Industry program. Most of his undergraduate teaching is devoted to project advising, both on- and off-campus, and to teaching introductory differential equations. He is the author of *Differential Equations: Modeling with Matlab* (Prentice Hall, 1999). He has consulted to industrial organizations ranging from small start-ups to large international firms. His current research is developing estimation and optimization algorithms to support operational and business decisions in deregulated utility markets. He began learning applied mathematics at Rensselaer Polytechnic Institute.

Wade Ellis, Jr. received his education at Phillips Andover Academy, Oberlin College, and The Ohio State University. He has taught mathematics at West Valley College for 24 years. While at West Valley College, he has been department chair, director of the mathematics and science computer laboratory, and faculty senate president. He has also taught at the University of California at Santa Barbara as a Visiting Research Professor and at the United States Military Academy as the Visiting Professor of Mathematics. He is a former member of the Mathematical Sciences Education Board and the Committee on Science Education K-12 of the National Research Council. He has served as a member of the Mathematical Association of America's Committee on Calculus Reform and the First Two Years. Wade has co-authored over 25 books on the use of technology in teaching and learning mathematics including the *Maple V Flight Manual*, *Calculus: Mathematics and Modeling*, and the *ODE Architect*.

Susan L. Ganter is Associate Professor of Mathematical Sciences at Clemson University. She has directed several local and national evaluation studies, including a recent residency at the National Science Foundation in which she investigated the national impact of the calculus reform initiative and helped to develop the evaluation plan for the Institution-wide Reform Program in the Division of Undergraduate Education. In addition, her work has included partnerships with industry that promote outreach to secondary mathematics students as well as professional development opportunities for secondary mathematics and science teachers. Dr. Ganter was formerly the Director of the

Program for the Promotion of Institutional Change at the American Association for Higher Education and a member of the Mathematical Sciences faculty at Worcester Polytechnic Institute.

Melvin D. George is President Emeritus of The University of Missouri, President Emeritus of St. Olaf College, and Professor Emeritus of Mathematics, University of Missouri-Columbia. After receiving a Ph.D. in Mathematics from Princeton University, George joined the faculty of the University of Missouri in 1960. He moved to the University of Nebraska-Lincoln in 1970 as Dean of the College of Arts and Sciences, returned to the University of Missouri as systemwide Vice President for Academic Affairs in 1975, and served as Interim President in 1984 before moving to St. Olaf College in Minnesota as President in 1985. Following his retirement from St. Olaf, George served for nearly two years as Vice President for Institutional Relations at the University of Minnesota. He returned to Missouri in 1996 and served once again as Interim President of the University of Missouri system in 1996–97. During 1994–96, he chaired the National Science Foundation's review of undergraduate science, mathematics, engineering, and technology education, culminating in the report, *Shaping the Future: New Expectations in Undergraduate Education in SMET.*

Sheldon P. Gordon is professor of mathematics at SUNY at Farmingdale. He was project director of the NSF-supported Math Modeling/PreCalculus Reform Project and principal author of the project text, *Functioning in the Real World: A PreCalculus Experience*. The project was recently awarded the top prize in the algebra through precalculus category in a competition sponsored by the Annenberg Foundation/Corporation for Public Broadcasting to identify exemplary Innovative Projects Using Technology. Gordon is a member of the Harvard Calculus Consortium that has produced the texts *Calculus*, *Multivariable Calculus*, *Applied Calculus*, and *Brief Calculus*. He is the co-editor of the MAA Notes volumes, *Statistics for the Twenty First Century* and *Calculus: The Dynamics of Change*, and is the author of over 90 articles. He is also a co-project director of the NSF-supported Long Island Consortium for Interconnected Learning in Quantitative Disciplines that seeks to connect mathematics to all other quantitative disciplines.

Harvey B. Keynes is Professor of Mathematics and Director of the Institute of Technology Center for Educational Programs at University of Minnesota. He was the past Director of Education at the NSF Geometry Center and has directed the University Talented Youth Mathematics Program (UMTYMP) for the past 20 years. Keynes was the leader of the Calculus Initiative (CI) development team and has extensively contributed to the progress of mathematics education for undergraduate and graduate students. These contributions include writing journal articles, participation in national and international conferences, directing special programs, and teaching professional development courses. He was the recipient of the 1992 American Mathematical Society Award for Distinguished Public Service.

Andrea Olson is Associate Director of the Institute of Technology Center for Educational Programs at University of Minnesota. Olson has a Ph.D. in Education Policy and Administration with an emphasis in faculty development. She was on the CI development team and manages program assessment, instructional personnel, and contributes to the grant writing and publications for pre-collegiate, undergraduate, and graduate mathematics and mathematics education.

Daniel J. O'Loughlin is a visiting Assistant Professor at Macalester College and was a Post-doctoral Fellow at University of Minnesota. O'Loughlin contributed to the development of the technologies used in the CI and taught the second year of the sequence. He holds four degrees in Mathematics or Mathematics Education, including a B.A. in Mathematics from St. John's University, Collegeville, MN and a Ph.D. in Mathematics from University of Minnesota.

Douglas Shaw is an Assistant Professor at Northern Iowa University and was a Post-doctoral Fellow at University of Minnesota. Shaw contributed to the CI instructional development component and to the curricular design of the group work. In addition, he taught both the first and second year of the sequence. He was chosen as the Best IT Instructor in the mathematics department at University of Minnesota for two years. Shaw received his Ph.D. from University of Michigan in 1995 and holds an undergraduate engineering degree.

James H. Lightbourne, III received his Ph.D. in Mathematics at North Carolina State University in 1976. His areas of research interest have been in partial differential equations, initially partial differential equations in abstract spaces and later work in multi-phase flow models. He joined the faculty at Pan American University in 1976 and at West Virginia University in 1979. Subsequently, he held a Visiting Research Position at the Mathematics Research Center of the University of Wisconsin-Madison. He served as Director of Graduate Studies of the WVU Department of Mathematics and became Chair of the Department in 1988. In 1991, he took a leave of absence from WVU to assume a rotator position in the NSF Division of Undergraduate Education to manage the NSF Calculus Program. In 1992, he accepted a permanent position at NSF as a Section Head in the Division. In his current position as Science Advisor, he has a primary responsibility in working on new initiatives and in cooperation with units throughout the Foundation.

William McCallum is a Professor of Mathematics at the University of Arizona. He received his Ph.D. in Mathematics from Harvard University in 1984, under the supervision of Barry Mazur. After spending two years at the University of California, Berkeley, and one at the Mathematical Sciences Research Institute in Berkeley, he joined the faculty at the University of Arizona in 1987. In 1989 he joined the Harvard Calculus Consortium. In 1993–94 he spent a year at the Institut des Hautes Etudes Scientifiques, and in 1995–96 he spent a year at the Institute for Advanced Study on a Centennial Fellowship from the American Mathematical Society. His professional interests include arithmetic algebraic geometry and renewal of the undergraduate curriculum, and he has written in both areas. He is a co-author of the Harvard Consortium calculus text and is the project director for the multi-variable calculus text. McCallum is currently involved in a curriculum development project on the mathematics of decision-making, which aims to develop realistic case studies for teaching mathematics to students in business and public administration.

David A. Smith received a Ph.D. from Yale University in 1963 and has been a member of the Duke University faculty since 1962. He has been on several editorial boards of journals and other publications concerning mathematics, computing, and education, and he has been active in the affairs of the Mathematical Association of America. He is the author of *Interface: Calculus and the Computer*, lead editor of *Computers and Mathematics: The Use of Computers in Undergraduate Instruction*, and co-author of *Calculus: Modeling and Application*. His published research papers have been in the areas of abstract algebra, combinatorics, mathematical psychology, numerical analysis, and mathematics education. Smith is Co-Director of *Project CALC: Calculus As a Laboratory Course* and the Duke component of the Connected Curriculum Project, both funded by the National Science Foundation.

Index